BOSTON STUDIES IN THE PHILOSOPHY OF SCIENCE

VOLUME LII

DIALECTICS OF THE CONCRETE

SYNTHESE LIBRARY

MONOGRAPHS ON EPISTEMOLOGY,

LOGIC, METHODOLOGY, PHILOSOPHY OF SCIENCE,

SOCIOLOGY OF SCIENCE AND OF KNOWLEDGE,

AND ON THE MATHEMATICAL METHODS OF

SOCIAL AND BEHAVIORAL SCIENCES

Managing Editor:

JAAKKO HINTIKKA, *Academy of Finland and Stanford University*

Editors:

ROBERT S. COHEN, *Boston University*

DONALD DAVIDSON, *Rockefeller University and Princeton University*

GABRIËL NUCHELMANS, *University of Leyden*

WESLEY C. SALMON, *University of Arizona*

VOLUME 106

BOSTON STUDIES IN THE PHILOSOPHY OF SCIENCE

EDITED BY ROBERT S. COHEN AND MARX W. WARTOFSKY

VOLUME LII

KAREL KOSÍK

DIALECTICS
OF THE CONCRETE

A Study on Problems of Man and World

D. REIDEL PUBLISHING COMPANY

DORDRECHT-HOLLAND / BOSTON-U.S.A.

Cloth edition: ISBN 90–277–0761–8
Paperback edition: ISBN 90–277–0764–2

DIALEKTIKA KONKRÉTNÍHO

Translated from the Czech by Karel Kovanda with James Schmidt

Published by D. Reidel Publishing Company,
P.O. Box 17, Dordrecht, Holland

Sold and distributed in the U.S.A., Canada and Mexico
by D. Reidel Publishing Company, Inc.,
Lincoln Building, 160 Old Derby Street, Hingham,
Mass. 02043, U.S.A.

EDITORIAL PREFACE

Kosík writes that the history of a text is in a certain sense the history of its interpretations. In the fifteen years that have passed since the first (Czech) edition of his *Dialectics of the Concrete*, this book has been widely read and interpreted throughout Europe, in diverse centers of scholarship as well as in private studies. A faithful English language edition is long overdue. This publication of Kosík's work will surely provoke a range of new interpretations. For its theme is the characterization of science and of rationality in the context of the social roots of science and the social critique which an appropriately rational science should afford.

Kosík's question is: How shall Karl Marx's understanding of science itself be understood? And how can it be further developed? In his treatment of the question of scientific rationality, Kosík drives bluntly into the issues of gravest human concern, not the least of which is how to avoid the pseudo-concrete, the pseudo-scientific, the pseudo-rational, the pseudo-historical. Starting with Marx's methodological approach, of "ascending from the abstract to the concrete", Kosík develops a critique of positivism, of phenomenalist empiricism, and of "metaphysical" rationalism, counterposing them to "dialectical rationalism". He takes the category of the concrete in the dialectical sense of that which comes to be known by the active transformation of nature and society by human purposive activity. In his wide-ranging critique of contemporary science and culture, Kosík gives a detailed account and interpretation of Marx's own methodology, in *Capital*.

Kosík's understanding of science, nature, human nature, and culture deserve a lively new audience with this translation, for the methodological and philosophical understanding of social science must once more try to come to terms with the genius of Karl Marx. Kosík's insights into the sciences are the outcome of his evident concern to read Marx once again, faithfully and deeply. May we, for our part, point briefly to Kosík on science?

> "The purely intellectual process of science transforms man into an abstract unit, integrated in . . . a system, (and) this reflects the real metamorphosis of man performed by capitalism".

> ". . . through the methodological approach, reality itself is changed: *methodology is ontologized*".

"Man can penetrate the mysteries of nature only because he *forms* a
human reality".
"Human praxis unites causality and purposiveness".
". . . cybernetics posed anew the question of what is specifically human".
". . . Marx proved this objective character of laws of science . . . independent of the scientist's subjective intentions".

Now we invite readers to think through Karel Kosík's understanding of
these provocative themes in the philosophy of the sciences, which lead to his
understanding of the concrete human life.

R. S. COHEN
M. W. WARTOFSKY

Center for Philosophy and History of Science,
Boston University
October 1976

TABLE OF CONTENTS

CHAPTER I

DIALECTICS OF THE CONCRETE
TOTALITY

THE WORLD OF THE PSEUDOCONCRETE AND ITS DESTRUCTION

Dialectics is after the 'thing itself'. But the 'thing itself' does not show itself to man immediately. To grasp it calls not only for a certain effort but also for a detour. Dialectical thinking therefore distinguishes between the idea of a thing and the concept of a thing, by which it understands not only two forms and two degrees of *cognition* of reality but above all two categories of human *praxis*. Man approaches reality primarily and immediately not as an abstract cognitive subject, as a contemplating head that treats reality speculatively, but rather as an objectively and practically acting being, an historical individual who conducts his practical activity related to nature and to other people and realizes his own ends and interests within a particular complex of social relations. As such, reality stands out to man not primarily as an object of intuition, investigation, and theorizing, whose opposite and complementary pole would be an abstract cognitive subject existing outside and beyond the world, but rather as the realm of his sensory—practical activity, which forms the basis for immediate practical intuition of reality. In his practical—utilitarian treatment of things, with reality appearing as the world of means, ends, tools, needs and procuring, the 'involved' individual forms his own ideas of things and develops an entire system of appropriate intuitions for capturing and fixing the phenomenal shape of reality.

'Real existence' and phenomenal forms of reality are directly reproduced in the minds of agents of historically determined praxis as a set of ideas or as categories of 'routine thinking' (considered only out of a 'barbarian habit' to be concepts). But these phenomenal forms are diverse and often contradict the *law* of the phenomenon, the *structure* of the thing, i.e., its *essential* inner kernel and the corresponding concept. People use money and carry out the most complicated transactions with it without ever knowing, or having to know, what money is. Immediate utilitarian praxis and corresponding routine thinking thus allow people to find their way about in the world, to feel familiar with things and to manipulate them, but it does not provide them with a *comprehension* of things and of reality. That is why Marx could have written that agents of social conditions feel at ease, as fish do in water, in the world of phenomenal forms that are alienated from their internal connections and are in such isolation absolutely senseless. They see nothing mysterious in what is through-and-through contradictory, and in their contemplation they take no exception to the inversion of the rational and the irrational. The praxis we are talking about here is the historically determined, one-sided and fragmentary praxis of individuals, based on the division of labor, the class differentiation of society and the resulting hierarchy of social status. What is formed in this praxis is both a particular material environment of the historical individual, and the spiritual atmosphere in which the superficial shape of reality comes to be fixed as the world of fictitious intimacy, familiarity and confidence within which man moves about 'naturally' and with which he has his daily dealings.

The collection of phenomena that crowd the everyday environment and the routine atmosphere of human life, and which penetrate the consciousness of acting individuals with a regularity, immediacy and self-evidence that lend them a semblance of autonomy and naturalness, constitutes the world of the *pseudoconcrete*. This world includes:

the world of external phenomena which are played out on the surface of real essential processes;

the world of procuring and manipulation, i.e., of man's fetishised praxis (which is not identical with the revolutionary-critical praxis of mankind);

the world of routine ideas which are external phenomena projected into man's consciousness, a product of fetishised praxis; they are ideological forms of the movement of this praxis;

the world of fixed objects which give the impression of being natural conditions and are not immediately recognizable as the result of man's social activity.

The world of the pseudoconcrete is the chiaroscuro of truth and deceit. It thrives in ambiguity. The phenomenon conceals the essence even as it reveals it. The essence manifests itself in the phenomenon, but only to a certain extent, partially, just in certain sides and aspects. The phenomenon indicates something other than itself and exists only thanks to its opposite. The essence is not immediately given: it is mediated by the phenomenon and thus shows itself in something other than what it is itself. The essence manifests itself in the phenomenon. Its manifestation in the phenomenon signifies its movement and proves that the essence is not inert and passive.

But the phenomenon similarly reveals the essence. Revealing the essence is the activity of the phenomenon.

The phenomenal world has its structure, its order and its laws that can be exposed and described. But the structure of the phenomenal world does not yet capture the relationship between this world and the essence. If the essence did not show itself in the phenomenal world at all, then the world of reality would be radically and fundamentally distinct from that of phenomena. The world of reality would be 'the other world' for man, as in Platonism or Christianity, and the only world accessible to him would be that of phenomena. But the phenomenal world is not something autonomous and absolute: phenomena turn into a phenomenal world while related to the essence. The phenomenon is not radically distinct from the essence, nor does the essence belong to a different order of reality. If this were the case, the phenomenon would have no internal relation to the essence; it could not reveal the essence while covering it up, their relationship would be one of mutual externality and indifference. To capture the phenomenon of a certain thing is to investigate and describe how the thing itself manifests itself in that phenomenon but also how it hides in it. Grasping the phenomenon negotiates *access* to the essence. Without the phenomenon, without this activity of manifesting and revealing, the essence itself would be beyond reach. In the world of the pseudoconcrete, the phenomenal aspect of the thing, in which the thing reveals and conceals itself, is considered to be properly the essence, and the distinction between the phenomenon and the essence *disappears*. Is thus the distinction between the phenomenon and the essence the same as between the real and the unreal, or as between two different orders of reality? Is the essence any more real than the phenomenon? Reality is the unity of the phenomenon and the essence. Consequently, the essence could be equally as unreal as the phenomenon, and vice-versa, *if* either one were isolated and in this isolation considered to be the one and only 'authentic' reality.

Thus the phenomenon is above all something that shows itself immediately, contrary to the concealed essence. But why does the 'thing itself', the structure of the thing, not show itself immediately and directly? Why must one undertake a detour and exert effort in order to grasp it? Why is the 'thing itself' concealed from immediate perception? In what way is it concealed? It cannot be concealed absolutely; for if man can at all *search* for the structure of the thing and if he wants to investigate this 'thing itself', if it is at all possible to expose the concealed essence or the structure of society, then prior to any investigation man already has to have a certain cognizance that there exists something such as the structure of the thing, the essence of the thing, the 'thing itself', that there exists a hidden truth of things which is different from phenomena that reveal themselves immediately. Man undertakes a detour and exerts an effort in exposing truth only because he somehow assumes that there is a truth to be exposed, and because he has a certain cognizance of the 'thing itself'. But why is the structure of the thing not accessible directly and immediately? Why is a detour necessary to capture it? And, where does the detour lead to? If the phenomenon of the thing is grasped in immediate perception, rather than the 'thing itself', is it because the structure of the thing is a reality of a different order than is the phenomenon? Is it consequently a different reality altogether, one that is *behind* phenomena?

The essence, unlike phenomena, does not manifest itself to us directly, and the concealed basis of things has to be *exposed in a specific activity.* This is precisely why science and philosophy exist. If the phenomenal form and the essence of things were coterminous, science and philosophy would be superfluous.[1]

Since ancient times, effort aimed at exposing the structure of things and the 'thing itself' has always been a matter for philosophy. Different *significant* philosophical trends are but so many variations of this basic problem and of solutions to it at different stages of the development of mankind. Philosophy is an *indispensable activity of mankind* because the essence of things, the structure of reality, the 'thing itself', the being of existents do not show themselves directly and immediately. In this sense, philosophy can be characterized as a systematic and critical effort directed at capturing the thing itself, at uncovering the structure of things, at exposing the being of existents.

The concept of the thing means comprehending the thing, and comprehending the thing means knowledge of the thing's structure. The most proper characteristic of cognition is its dividing the one. Dialectics does not enter cognition from without or as an afterthought, nor is it a property of cognition. Rather, cognition is dialectics itself, in one of its forms: cognition is dividing the one. In dialectical thinking, the terms 'concept' and 'abstraction' have the significance of a method that divides the one in order to intellectually reproduce the structure of the thing, i.e., to comprehend it.[2]

Cognition is realized as separation of the phenomenon from the essence, of the peripheral from the essential, because only such a separation can demonstrate their internal connection and thus the specific character of the thing. In this process, the peripheral is not cast aside, it is not separated out as less real or as unreal. Instead, its character is demonstrated as being phenomenal or peripheral by proving the truth of the thing in its essence. This division of the one which is a constitutive element of philosophical cognition – *there is no cognition without division* – displays a structure analogous to that of human activity: for activity, too, is based on dividing the one.

The fact that thinking spontaneously moves in a direction counter to the character of reality, that it has an isolating and 'paralysing' effect, and that this spontaneous movement contains a tendency toward abstractness, is not in itself an immanent property of thinking, but rather follows from its practical function. All activity is 'one-sided'[3] because it pursues a *particular* goal, and therefore isolates some moments of reality as essential while leaving others aside. This spontaneous activity elevates certain moments important for attaining particular goals and thus cleaves a unified reality, intervenes in reality, 'evaluates' reality.

The spontaneous inclination of 'praxis' and thinking to isolate phenomena and to divide reality into what is essential and what is peripheral is *always* accompanied by an *awareness of the whole* in which and from which certain aspects have been isolated. This awareness is also spontaneous, though it is less clearly apparent to naive consciousness, and is frequently unconscious. Dim awareness of a 'horizon of indeterminate reality' *as a whole* is the ubiquitous *backdrop* of all activity and thinking, unconscious though it may be for naive consciousness.

Phenomena and phenomenal forms of things are spontaneously reproduced in routine thinking as reality (i.e., as reality itself) not because they are on the surface and thus closest to sensory cognition, but because the phenomenal form of things is the natural product of everyday praxis. The everyday utilitarian praxis gives rise to 'routine thinking' – which covers both familiarity with things and with their superficial appearance, and the technique of handling things in practice – as a form of movement and existence. But the world that exposes itself to man in his fetishised praxis, in procuring and manipulation, is not a real world, though it does have a real world's 'firmness' and its 'effectiveness'; rather, it is a 'world of appearances' (Marx). The idea of a thing postures as the thing itself and forms an ideological appearance but it is not a natural property of things and of reality; rather, it is the projection of certain *petrified* historical conditions into the consciousness of the subject.

Distinguishing between the idea and the concept, between the world of

appearances and that of reality, between everyday utilitarian praxis of
people and the revolutionary praxis of mankind, in one phrase: 'dividing the
one', is the mode by which thinking penetrates to the 'thing itself'.
Dialectics is critical thinking that strives to grasp the 'thing itself' and
systematically searches for a way to grasp reality. Dialectics is thus the
opposite of doctrinaire systematization or romanticization of routine ideas.
Thinking that wants to know reality adequately will be satisfied neither
with abstract schemes of this reality nor with equally abstract ideas of it. It
therefore has to *abolish** the apparent autonomy of the world of immediate
everyday contacts. Such thinking, which abolishes the pseudoconcrete in
order to reach the concrete, is also a process that exposes a real world under
the world of appearances, the law of the phenomenon behind the
appearance of the phenomenon, real internal movement behind the visible
movement, the essence behind the phenomenon.[4] What lends these
phenomena a pseudoconcrete character is not their existence as such but the
apparent autonomy of their existence. In destroying the pseudoconcrete,
dialectical thinking does not deny the existence or the objective character of
these phenomena, but rather abolishes their fictitious independence by
demonstrating their mediatedness, and counters their claim to autonomy
with proving their derivative character.

Dialectics does not consider fixed artifacts, formations and objects, the
entire complex of both the material world of things and that of ideas and of
routine thinking, to be something original and autonomous. It does not
accept them in their ready-made form, but subjects them to investigation in
which the reified forms of the objective and the ideal worlds dissolve, lose
their fixed and natural character and their fictitious originality, and show up
as derivative and mediated phenomena, as sediments and artifacts of the
social praxis of mankind.[5]

Uncritical reflective thinking[6] will immediately, i.e., with no dialectical
analysis, causally relate fixed ideas with equally fixed conditions, and will
present this manner of 'barbarian thinking' as a 'materialist' analysis of
ideas. Since people have been aware of their own time (i.e., they have
experienced, evaluated, criticised and grasped it) in categories of 'the
collier's faith' or of 'petit-bourgeois scepticism', the doctrinaire believes that
he has 'scientifically' analysed these ideas once he identifies their corresponding economic, social, or class equivalents. This 'materialization' of
course accomplishes nothing but a double mystification: the inversion of

*See note on p. 99.

the world of appearances (of fixed ideas) is anchored in an inverted (reified) materiality. Marxist theory has to *initiate* the analysis by asking *why* were people aware of their own time precisely in these categories, and *what kind of a time* do people find reflected in them. With this question, the materialist prepares the ground for destroying the pseudoconcrete *both* of ideas *and* of conditions, *whereupon* he can suggest a rational explanation of the internal connection between the times and the ideas.

The destruction of the pseudoconcrete, the dialectical—critical method of thinking that dissolves fetishised artifacts both of the world of things and of that of ideas, *in order* to penetrate to their reality, is of course only another aspect of dialectics as a *revolutionary* method of *transforming reality*. *To interpret the world critically, the interpretation itself must be grounded in revolutionary praxis.* We shall see later on that reality can be transformed in a *revolutionary* way only because, and only insofar as, we ourselves form reality, and know that reality is formed by us. In this respect, the difference between natural reality and socio-human reality is this, that though man can change and transform nature, he can change socio-human reality *in a revolutionary way*; but he can do so only because he forms *this* reality himself.

The real world, concealed by the pseudoconcrete, and yet manifesting itself in it, is neither a world of real conditions opposed to unreal ones, nor a world of transcendence opposed to a subjective illusion, but a world of human praxis. It is the comprehension of socio-human reality as the *unity* of production and products, of subject and object, of genesis and structure. The real world is thus not the world of fixed 'real' objects leading a transcendental existence behind their fetishised forms, as in some naturalistic parallel to Platonic ideas; rather, it is a world in which things, meanings and relations are conceived as *products* of social man, with man himself exposed as the real subject of the social world. The world of reality is not a secularized image of paradise, of a ready-made and timeless state, but is a process in which mankind and the individual *realize* their truth, i.e., humanize man. The world of reality, unlike the world of the pseudoconcrete, is a world of *realizing* truth, a world in which truth is not given and preordained, and as such copied, ready-made and immutable, in human consciousness, but rather a world in which truth *happens*. This is why human history can be the story of truth and the happening of truth. Destroying the pseudoconcrete means that truth is neither unattainable, nor attainable once and for all time, but that truth itself happens, i.e., develops and realizes itself.

The pseudoconcrete is thus destroyed in the following ways: (1) by the revolutionary–critical praxis of mankind which is identical with the humanization of man, with social revolutions as its key stages; (2) by dialectical thinking which dissolves the fetishised world of appearances in order to penetrate to reality and to the 'thing itself'; (3) by the realization of truth and the forming of human reality in an ontogenetic process; since the world of truth is also the own individual creation of every human individual as a social being. Every individual has to appropriate his own culture and lead his own life *by himself and non-vicariously.*

Destroying the pseudoconcrete is thus not like tearing down a curtain to discover a ready-made and given reality, existing independently of man's activity hiding behind it. The pseudoconcrete is precisely the autonomous existence of man's *products* and the reduction of man to the level of utilitarian praxis. Destroying the pseudoconcrete is the process of forming a concrete reality and of seeing reality in its concreteness. Idealist trends have either absolutized the subject, and deal with the problem of how to look at reality so that it be concrete and beautiful, or they have absolutized the object, and believe that the more perfectly the subject is eliminated from reality, the more real reality is. The materialist destruction of the pseudoconcrete by contrast results in the liberation of the 'subject' (i.e., in concrete *seeing* of reality as opposed to fetishist 'intuiting' of it) merging with the liberation of the 'object' (with the forming of a human environment in terms of humanly transparent and rational conditions), because the social reality of people forms itself as a dialectical unity of the subject and the object.

The call 'ad fontes' that one periodically hears as a reaction *against* the most diverse manifestations of the pseudoconcrete, as well as the positivist methodological rule of 'presuppositionlessness', have their basis and substantiation in the materialist destruction of the pseudoconcrete. The return to 'the sources' takes on two entirely different forms, though. At times it appears as a humanist, scholarly, learned critique of sources, as an investigation of archives and of antiquities, from which true reality is to be derived. But in its more profound and more important form, which even learned scholasticism finds barbaric (as testified by reactions to Shakespeare and Rousseau), the call 'ad fontes' signifies a *critique of civilization and culture*, a romantic or a revolutionary attempt to discover productive activity behind products and artifacts, to find the 'real reality' of the concrete man behind the reified reality of reigning culture, to dig out the authentic subject of history from under the sediment of fixed conventions.

THE SPIRITUAL AND INTELLECTUAL REPRODUCTION OF REALITY

Because things do not show man immediately what they are, and because man does not have the ability to immediately intuit things in their essence, mankind arrives at the cognition of things and of their structure via a detour. Precisely because this detour is the *only* negotiable path to truth, every now and then will mankind attempt to spare itself the trouble of the long journey and seek to intuit the essence of things *directly* (mysticism is man's impatience in the search for truth). But man is also in danger of losing his way on this detour, or of getting stuck halfway.

'Self-evidence,' far from being the evidence and clarity of the thing itself, is the opacity of the idea of the thing. What is natural shows up as unnatural. Man has to exert effort to emerge from his 'state of nature' and to become a man (man *works himself up* to being a man) and to recognize reality for what it is. Great philosophers of all times and tendencies, Plato with his myth of the cave, Bacon with his image of idols, Spinoza, Hegel, Husserl and Marx, have all correctly characterized cognition as overcoming that which is natural, as supreme activity and 'use of force'. The dialectic of activity and passivity in human cognition is manifest particularly in the fact that in order to know things in themselves, man has to transform them into things for himself; to know things as they are independently of him, he has to subject them to his praxis; to find out how they are without his interference he has to interfere with them. Cognition is not contemplation. Contemplation of the world is based on the results of human praxis. Man knows reality only insofar as he *forms* a human *reality* and acts primordially as a practical being.

In order to come close to the thing and its structure, and to find access to it, some distance is imperative. It is well known how *difficult* it is to deal scientifically with current events, whereas analysing events past is relatively easier, for reality itself has performed a certain elimination or a 'critique'. Science has to replicate this natural course of history *artificially* and experimentally. What is the basis of this experiment? It is the appropriate and substantiated *distance* of science, from which things and events are seen adequately and without distortion. (The importance of this thought experiment which substitutes for real historical distance has been emphasized by Schiller, in the context of drama.)

The structure of the thing, that is, the thing itself, can be grasped neither immediately, nor by contemplation or mere reflection, but only by a certain *activity*. It is impossible to penetrate to the 'thing itself' or to answer the question, what the 'thing itself' is, without analysing the activity through which the thing is grasped. Such an analysis has to cover also the problem of *creating* this very activity which negotiates access to the 'thing itself'. These activities are different kinds or modes of human *appropriation* of the world. Problems elaborated in phenomenology under such descriptions as 'intentionality toward something', 'intention of meaning toward something', or as various 'modes of perception' have been interpreted on a materialist basis by Marx, as various kinds of human *appropriation* of the world: the spiritual-practical, theoretical, artistic, religious, but also the mathematical, physical, etc. One cannot appropriate mathematics, and thus grasp it, with an intentionality that is not appropriate for mathematical reality, e.g. with a religious experience or with artistic perception. Man lives in several worlds, but to each of them there is a different key. One cannot move from one world to another without the right key, i.e. without changing the intentionality and the mode of appropriating reality. In modern philosophy and modern science, which have been permanently enriched by the concept of praxis, *cognition* represents one mode of man's *appropriating* the world; and *every* such mode of appropriation has two constitutive elements, namely its subjective and its objective *sense*. What is the intentionality, what is the view, the *sense* that man has to develop, to 'rig up', in order to grasp and uncover the *objective sense* of the thing? The process of capturing and exposing the *sense* of the thing amounts at the same time to forming the appropriate 'sense' in man with which he can comprehend the sense of the thing. The *objective sense* of the thing can be grasped if man cultivates the *appropriate sense*. These senses with which man uncovers both reality and the sense of reality are themselves an historical–social product.[7]

All degrees of human cognition, sensory or rational, as well as all modes of appropriating reality, are activities based on the objective praxis of mankind, and are consequently in some degree connected with and in some way mediated by all other modes. Man always perceives *more* than what he sees and hears immediately. The building that I see in front of me I perceive primordially and immediately as an apartment house, a factory or as an historical monument, and this immediate sensory perception is realized in a certain mood which manifests itself as interest, indifference, astonishment, revulsion, etc. In the same way, the din I hear, I perceive first of all as the din of an approaching or departing plane, and I can tell by the very sound whether it is a 'copter, jet, fighter or transport plane, etc. Thus in a certain way, *all* of my knowledge and culture participates in my hearing and seeing, as do all my experiences, current or those buried in oblivion to be recovered in certain situations, and all my thinking and judgement, although none of this manifests itself in concrete acts of perceiving and experiencing in any explicitly predicative form. Thus in the course of appropriating the world spiritually–practically, which is the basis for all other modes of appropriation – the theoretical, artistic, etc. – reality is perceived as an *undifferentiated whole of existents and of meanings*, and it is implicitly grasped in a unity of statements of fact and those of value. It takes abstraction and thematization, a *project*, to select out of this full and inexhaustible world of reality certain areas, aspects and spheres, which naive naturalism and positivism would then consider to be the *only* true ones and the only reality, while suppressing the 'rest' as sheer subjectivity. The physicalist image presented by positivism impoverishes the human world, and its absolute exclusiveness deforms reality, because it reduces the real world to but *one* dimension and aspect, to the dimension of extensity and of quantitative relations. In addition, it cleaves the human world, when it declares the world of physicalism, the world of idealised real values, of extensity, quantity, mensuration and geometric shapes to be the only reality, while calling man's everyday world a fiction.

In the world of physicalism that modern positivism considers to be the only reality, man can exist only in a particular abstract activity, i.e. as a physicist, statistician, mathematician, or a linguist, but not in all of his potentialities, not as a whole man. The physical world, a thematized mode of cognition of the physical reality, is only one of the possible images of the world, and expresses certain essential properties and aspects of objective reality. Apart from the physical world there exist other worlds, too, and equally *justified* ones: e.g., the artistic, the biological, etc.; in other words, reality *is not exhausted* in the physical picture of the world. Positivist physicalism has substituted a certain *image* of reality for *reality* itself and has promoted a certain mode of appropriating the world as the only true one. Thereby it denied, first, the inexhaustibility of the objective world and its irreducibility to knowledge, which is one of the fundamental theses of materialism, and, second, it impoverished the human world by reducing the wealth of human subjectivity, formed *historically* through the objective praxis of mankind, to one single mode of appropriating reality.

Every particular thing upon which man focuses his view, attention, action or evaluation, emerges from a certain whole which envelops it and which man perceives as an indistinct background or as a dimly intuited imaginary context. How does man perceive individual things? As absolutely isolated and unique, perhaps? Actually, he *always* perceives them in a horizon of a certain *whole*, which is usually unexpressed and not perceived explicitly. Whatever man perceives, observes, works on, is a part of a whole, and it is precisely this not explicitly perceived whole which is the light that illuminates and reveals the very uniqueness and significance of the unique thing under observation. Human consciousness therefore has to be invest-igated both in its theoretical—predicative form, of explicit, substantiated rational and theoretical cognition, and in its pre-predicative, holistically intuitive form. Consciousness is the unity of both forms which intermingle and influence one another, because they are based, united, on objective praxis and on the spiritual—practical reproduction of reality. Denying or invalidating the first form leads to irrationalism and to assorted varieties of 'vegetative thinking', whereas denying or underrating the second form leads to rationalism, positivism and scientism which in their one-sidedness inexorably produce irrationality as their own complement.

Yet why does theoretical thinking turn into a 'universal medium' through which everything that had been experienced in an experience, intuited in an intuition, imagined in an idea, performed in an action and felt in a feeling has to *once again* make its passage? Why is the reality which man appropriates above all spiritually—practically, and on this basis also artistically, religiously, etc., the reality that man experiences, evaluates, and works on, why is it appropriated *once again* theoretically? A certain 'privileged character' of the theoretical sphere over all others can be demonstrated in the fact that anything can become a topic for theory and subjected to explicit analytical investigation: aside from art there is a theory of art, aside from sport there is a theory of sport, aside from praxis a theory of praxis. What is this 'privileged character' about? Does perhaps the truth of art lie in the theory of art, and the truth of praxis in the theory of praxis? Does the impact of art follow from the theory of art and the impact of praxis from its own particular theory? These are indeed the assumptions of every caricature and of every formalist—bureaucratic concept of theory. Theory, however, determines neither the truth nor the impact of this or that non-theoretical kind of appropriating reality, but represents rather the *explicitly reproduced* comprehension of the corresponding kind of appropriating, whose intensity, truthfulness, etc. it influences in its own turn.

Materialist epistemology, as the spiritual reproduction of society, captures the *two-fold* character of consciousness which both positivism and idealism miss. Human consciousness is at once a 'reflection' and a 'project', it registers as well as constructs and plans, it both reflects and anticipates, is both receptive and active. To let the 'thing itself' express itself, to add nothing and just let things be as they are — this requires a special activity.

Epistemology as the spiritual reproduction of society emphasizes the active character of cognition on *all* levels. Elementary sensory knowledge is not the result of passive perception but of perceptional activity. Yet, as incidentally follows from the central tenet of this work, every epistemology is implicitly or explicitly based on a certain theory of reality, and presupposes a certain concept of reality. Materialist epistemology, as the intellectual reproduction of society, is based on a conception of reality different from that of the method of reduction. Reduction presupposes a rigid substance and immutable, further irreducible elements, to which the diversity and variety of phenomena can in the last analysis be reduced. The phenomenon is considered explained when reduced to its essence, to a general law, to an abstract principle. How untenable reductionism is for social reality has been demonstrated by a well-known observation: Franz Kafka is a petit-bourgeois intellectual; yet not every petit-bourgeois intellectual is a Franz Kafka. The method of reductionism subsumes the unique under the generally abstract, and posits two unmediated poles: abstract individuality on the one end and abstract generality on the other.

Spinozism and physicalism are the two most wide-spread varieties of the reductionist method which translates the wealth of reality into something basic and elementary. All the richness of the world is jettisoned into the abyss of an immutable substance. For Spinoza, this method is just another side of moral asceticism which proves that all wealth is actually non-wealth, that everything concrete and unique is illusory. There is a certain intellectual tradition that would consider Marx's theory to be dynamized Spinozism; as though Spinoza's immutable substance were set in motion. In this form, modern materialism would be of course merely a variation on metaphysics. Modern materialism has not dynamised an immutable substance, but has posited the 'dynamics' and the dialectics of being as the 'substance'. Coming to know the substance thus does not amount to reducing the 'phenomenon' to a dynamized substance, i.e. to something concealed behind phenomena as something independent of them; rather, it is cognition of the laws of movement of the thing itself. The very *movement of the thing, or the thing in motion*, is the 'substance'. The movement of the thing forms particular phases, forms and aspects that cannot be comprehended by reducing them to a substance, but that are comprehensible as an explication of the 'thing itself'. Religion can be materialistically comprehended not by finding the earthly kernel of religious artifacts or by reducing them to material conditions, but only as an inverted and mystified activity of man, the objective subject. The 'substance' of man is objective activity (praxis), not some dynamized substance in man.

Reductionism is the method of 'nothing but'. The wealth of the world is 'nothing but' a substance, immutable or dynamized. Therefore reductionism cannot rationally explain *new* phenomena, or qualitative development. It will reduce anything new to conditions and prerequisites; the new is 'nothing but' – the old.[8]

If the entire richness of man as a social being were reduced to the statement that the essence of man is the production of tools, and if the entire social reality were in the last analysis determined by economics, *in the sense of the economic factor*, the following question would arise: Why does this factor have to be disguised, why does it realize itself in forms that are innately alien to it, such as imagination and poetry?[9]

How can the new be comprehended? According to the above conception, by reducing it to the old, to conditions and prerequisites. New appears here as something external, as a supplement to material reality. Matter is in motion but does not have the property of negativity.[10] Only such a concept of matter that in matter itself discovers negativity, that is, the potentiality to produce new qualities and higher stages of development, can material-istically explain the new as a property of the material world. Once matter is grasped as negativity, scientific explanation no longer amounts to reduction, to reducing the new to prerequisites, to reducing concrete phenomena to an abstract base, and it instead becomes the *explication of phenomena*. Reality is explained not by reducing it to something other than what it is itself, but by having it explicate itself, in unfolding and illuminating its phases and aspects of its movement.[11]

The starting point of the investigation must be formally identical with the result. The identity of this starting point must be maintained throughout the whole course of thinking, as the only guarantee that thinking will not start its journey with Virginia Woolf and end it with the Big Bad Wolf. But the sense of the investigation is in this, that in a spiral movement, it reaches a result which had not been known at the outset, and thus that while the starting point and the result are formally identical, thinking does in the end arrive at something different in content than what it had started with. Thinking progresses from a vibrant, chaotic, immediate idea of the whole toward concepts, to abstract conceptual determinations, and in summing them up it returns to the starting point which no longer is an un-comprehended though vibrant whole of immediate perception, but a richly differentiated and comprehended whole of the concept. The journey from the 'chaotic idea of the whole' to the 'rich totality of many determinations and relations' is identical with comprehending reality. The whole is not cognizable by man immediately, though it is given immediately to his senses as the idea, the intuition, the experience. The whole that is immediately accessible to man is a chaotic and opaque whole. A detour is necessary in order to know and comprehend this whole, to clarify and explicate it: the concrete is comprehensible by way of the abstract, the whole by way of its parts. Precisely because the journey of truth is roundabout – *der Weg der Wahrheit ist Umweg* – man can lose his way or get stuck halfway.

The method of ascending from the abstract to the concrete is a method of *thinking*, in other words, it is a movement realized in the concepts and the life-element of abstraction. Ascending from the abstract to the concrete is not a transition from one level (the sensory) to another (the rational); it is rather movement in thinking and the motion of thought. If thinking is to ascend from the abstract to the concrete, it has to move in its own life-element, i.e. on an abstract level which is the negation of sensory immediacy, clarity and concreteness. Ascending from the abstract to the concrete is a movement for which every beginning is abstract and whose dialectics consists of transcending this abstractness. Ascending from the abstract to the concrete is therefore generally a movement from the part to the whole and from the whole to its parts, from the phenomenon to the essence and from the essence to the phenomenon, from totality to contradiction and from contradiction to totality, from the object to the subject and from the subject to the object. Ascending from the abstract to the concrete, which amounts to materialist epistemology, is the dialectics of the concrete totality in which reality is intellectually reproduced *on all levels and in all dimensions*. The process of thinking not only transforms the chaotic whole of ideas into a clear whole of concepts; but in this process, the whole itself is outlined, determined and comprehended, too.

As we know, Marx distinguished between the method of investigation and that of exposition. Nevertheless, the method of investigation is frequently passed over as something familiar, whereas the method of exposition is taken merely for a form of presentation. It is ignored that precisely this method renders the phenomenon transparent, rational and comprehensible. The method of investigation involves three stages:

(1) Appropriating the material in detail, mastering it to the last historically accessible detail.

(2) Analysing its different forms of development.

(3) Tracing out their internal connections, i.e. determining the unity of different forms in the development of the material.[12]

Without mastering this method of *investigation*, any dialectics is but barren speculation.

That with which science *initiates* its exposition is already the *result* of research and of a critical-scientific appropriation of the subject-matter. The beginning of the presentation is a *mediated* beginning which like a germ contains the construction of the whole work. But precisely what can and should serve as *the beginning of the exposition*, i.e. of the scientific unfolding (explication) of the problematique, is not known at *the beginning of the investigation*. The beginning of the exposition and the beginning of the investigation are two different things. The beginning of the investigation is random and arbitrary, the beginning of the exposition is necessary.

Marx's *Capital* begins — and this fact has since become trivial — by an analysis of a commodity. But the knowledge that a commodity is a cell of the capitalist society, an abstract beginning whose unfolding will reproduce the whole internal structure of the capitalist society — this origin of the *exposition* results from an investigation, from a scientific appropriation of the subject-matter. A commodity is an 'absolute reality' for the capitalist society because it is the unity of all determinations, the germ of all contradictions, and as such can be characterized in Hegelian terms as the unity of being and not-being, of the differentiated and the undifferentiated, of identity and non-identity. All other determinations are but richer definitions and concretizations of this 'absolute' of the capitalist society. The dialectics of the exposition or of the explication may not overshadow the central problem: how does science arrive at the *necessary origin of the presentation*, i.e. of the explication? Not distinguishing or indeed confusing the beginning of the investigation with that of the exposition (explication) in interpreting Marx's work becomes a source of the trivial and of the ridiculous. The beginning of the investigation is arbitrary but the presentation is an *explication of the thing* precisely because it presents the thing in its *necessary internal* development and unfolding. Here, the true beginning is the necessary beginning, and other determinations of necessity stem from it. Without a necessary beginning, the exposition is no unfolding, no explication, but mere eclectic accumulation or skipping from one thing to another, or finally, it is not the necessary internal unfolding of the *thing itself* but only an unfolding of the reflection of the thing, of the contemplation of the thing, which in relation to the thing itself is an external and arbitrary matter. The method of explication is no evolutionist unravelling, but rather the unfolding, exposing and 'complicating' of contradictions, the unfolding of the thing by way of contradictions.

Explication is a method that proves the unfolding of the thing to be a *necessary transformation* of the abstract into the concrete. Ignorance of the method of dialectical explication based on comprehending reality as a concrete totality leads either to subsuming the concrete under the abstract, or to skipping intermediate links and to creating forced abstractions.

Materialist dialectics as a method of scientific clarification of the socio—human reality thus is not a search for the earthly kernel of spiritual artifacts (as Feuerbach's reductionist, Spinozist materialism would have it), nor does it assign cultural phenomena to their economic equivalents (as Plekhanov had taught, in the same Spinozist tradition), or reduce culture to the economic factor. *Dialectics is not a method of reduction, but a method of spiritual and intellectual reproduction of society*, a method of unfolding and explicating social phenomena on the basis of the objective activity of the historical man.

CONCRETE TOTALITY

The category of totality, anticipated in modern thinking especially by Spinoza with his *natura naturans* and *natura naturata*, has been elaborated in German classical philosophy as a central concept for polemically distinguishing dialectics from metaphysics. The standpoint of totality, which grasps reality in its internal laws and uncovers necessary internal connections under superficial and haphazard phenomena, is juxtaposed against the standpoint of empiricism that dwells on such haphazard phenomena and cannot arrive at a comprehension of the development of reality. By the standpoint of totality we understand the dialectics of lawfulness and randomness, of parts and the whole, of products and producing, etc. Marx[13] adopted this dialectical concept, scoured it of its ideological mystifications and turned its *new* form into one of the central concepts of *materialist dialectics*.

But a strange fate befalls central concepts of philosophy, concepts which expose essential aspects of reality. They *always* cease to be the exclusive property of the philosophy which first employed and substantiated them, and they gradually move into the public domain. As a concept expands, as it becomes accepted and achieves general recognition, it undergoes a metamorphosis. The category of totality has also been well received and broadly recognized in the twentieth century, but it is in constant danger of being grasped one-sidedly, of turning into its very opposite and ceasing to be a *dialectical* concept. The main modification of the concept of totality has been its reduction to a *methodological* precept, a methodological rule for *investigating* reality. This degeneration has resulted in two ultimate trivialities: that everything is connected with everything else, and that the whole is more than the sum of its parts.

In materialist philosophy, the category of concrete totality answers first and foremost the question, *what is reality*. Only secondarily, and only after having materialistically answered the first question, can it be an epistemological principle and a methodological precept. Idealist trends of the 20th century have abolished the three-dimensionality of totality as a methodological principle and have reduced it to a single dimension — the relation of the whole to its parts.[14] In particular, though, they have radically severed totality as a methodological precept and an epistemological principle of the *cognition* of reality from the materialist conception for which *reality itself* is a concrete totality. Thus severed, totality can no longer be substantiated as a coherent methodological principle. It will instead be interpreted idealistically and its content will be impoverished.

Cognition of reality, its mode and its possibility, depend in the last analysis on an explicit or implicit conception of reality. The question, how can reality be known, is always preceded by a more fundamental question: What is reality?

What is reality, indeed? If it were only a sum of facts, of the simplest and further irreducible elements, then it would follow that, first, concreteness is the sum of *all* facts, and that, second, reality in its concreteness is principally unknowable because to every phenomenon one can array further facets and aspects, further forgotten or as yet undiscovered facts, and by this *infinite arraying* prove the abstract and inconcrete character of cognition. 'All knowledge, whether intuitive or discursive', notes a leading contemporary opponent of the philosophy of concrete totality, 'must be of abstract aspects, and we can never grasp the 'concrete structure of [social] reality itself'.[15]

There is a principal difference between the opinion that considers reality to be a concrete totality, i.e. a structural, evolving, self-forming whole, and the position that human cognition can, or cannot, achieve a 'totality' of aspects and facts, i.e. of *all* properties, things, relations and processes of reality. The second position takes totality as a sum of all facts. Since human cognition never can, in principle, encompass all facts, for additional facts and aspects can always turn up, this position considers the standpoint of concreteness or totality to be mysticism.[16] Totality indeed does not signify *all facts*. Totality signifies reality as a structured dialectical whole, within which and from which *any particular* fact (or any group or set of facts) can be rationally comprehended. The accumulation of all facts would not yet amount to the cognition of reality, and neither would all accumulated facts amount to a totality. Facts are the cognition of reality only provided they are comprehended as facts and as structural parts of a dialectical whole, i.e. not as immutable, further irreducible atoms which, agglomerated, compose reality. The concrete, *that is*, totality, is thus not equal to all the facts, to a sum of facts or to the accumulation of all aspects, things and relations, for this set lacks the most important feature — totality and concreteness. Without comprehending *what* facts signify, i.e. without comprehending that *reality* is a concrete totality which for the purposes of knowing individual facts or sets of facts *turns into* a structure of meanings, cognition of the concrete reality itself amounts to no more than mysticism or to a thing in itself unknowable.

The dialectics of the concrete totality is not a method that would naively aspire to know all aspects of reality exhaustively and to present a 'total' image of reality, with all its infinite aspects and properties. Concrete totality is not a method for capturing and describing *all* aspects, features, properties, relations and processes of reality. Rather, it is a theory of reality as a concrete totality. This conception of reality, of reality as concreteness, as a whole that is structured (and thus is not chaotic), that evolves (and thus is not immutable and given once and for all), and that is in the process of forming (and thus is not ready-made in its whole, with only its parts, or their ordering, subject to change), has certain methodological implications that will become a heuristic guide and an epistemological principle for the study, description, comprehension, interpretation and evaluation of certain thematic sections of reality, be it physics or literary criticism, biology or political economy, theoretical problems of mathematics or practical issues of organizing human life and social conditions.

In modern times, man's thinking has been leading to a dialectics of cognition, to a dialectical concept of cognition, which manifests itself especially in the dialectical relation of the absolute and the relative truth, the rational and the empirical, the abstract and the concrete, the premise and the conclusion, the assumption and the proof, etc. It has also, however, been leading to a comprehension of the dialectics of objective reality itself. The possibilities of creating a unified science and a unified concept of science are based on the exposition of a more profound unity of objective reality. The development of science in the 20th century has been noteworthy in that the more specialized and differentiated it becomes, and the more new areas it uncovers and describes, the more clearly evident is the *internal* material unity of most diverse and distant areas. This in turn leads to a fresh questioning of the relationships of mechanism and organism, of causality and teleology, etc., and thus also of the *unity of the world*. The differentiation of science at one point seemed to jeopardize the unity of science. It contained the danger of parcelling out the world, nature and matter into independent, isolated units, and of transforming scientists into isolated pilgrims in their own disciplines, each working out of context and deprived of means of communication. In fact, though, it has led to results and consequences which *actually* further a more profound exposition and cognition of the *unity* of reality. This profound comprehension of the unity of reality has its counterpart in an equally profound comprehension of the *specificity* of various areas and phenomena as well. In sharp contradiction to the romantic disdain for natural sciences and technology, it was precisely modern technology, cybernetics, physics and biology that have highlighted new potential for the development of humanism and for investigating that which is specifically human.

Attempts to create a new unified science stem from finding that the structure of reality itself is dialectical. The existence of structural similarities in areas that are quite diverse and internally quite different is based on the fact that all areas of objective reality are systems, i.e. complexes of interdependent elements.

The parallel development of different scientific disciplines, especially of biology, physics, chemistry, cybernetics and psychology, highlights the problem of organization, structure, wholeness, dynamic interaction, and leads to the recognition that the study of isolated parts and processes is insufficient. The main problem is 'organizing relations that result from dynamic interaction and make the behavior of parts different, when studied in isolation or within the whole'.[17] Structural similarities form a starting point for a more profound investigation of the *specificity* of phenomena. Positivism has conducted a grandiose purification of philosophy from remnants of the *theological* conception of reality, as a hierarchy of degrees of perfection. As the ultimate leveler it has reduced all reality to physical reality. The one-sidedness of the scientistic conception of philosophy should not overshadow the creditable destructive and demystifying role of modern positivism. Hierarchizing reality on a non-theological principle is possible only on the basis of degrees of *complexity* of structure and of forms of movement of reality itself. Hierarchizing systems on the basis of the complexity of their internal structure fruitfully continues in the tradition of

Enlightenment and in the heritage of Hegel who had also examined reality (which he conceived as a system) on this basis, describing internal structure in terms of mechanism, chemism and organism. But only the dialectical conception of the ontological and gnoseological aspects of structure and system provides a fruitful solution and avoids the extremes of mathematical formalism on the one side and of metaphysical ontologism on the other side. Structural similarities of various forms of human relations (language, economics, kinship patterns, etc.) can lead to a more profound understanding and explanation of social reality only as long as both the *structural similarities* and the *specificity* of these phenomena are respected.

The dialectical conception of the relationship between ontology and gnoseology allows one to detect the disparity and poor fit between the logical structure (model), used to interpret reality or some area of it, and the structure of this reality itself. A certain model, structurally of a 'lower order' than the corresponding area of reality, can interpret this more complex reality only approximately; the model can become the first approximation of an adequate description and interpretation. Beyond the limits of this first approximation, the interpretation is false. The concept of mechanism will, for example, explain the mechanism of a timepiece, the mechanism of memory, and the mechanism of social life (the state, social relations, etc.). But *only* in the first instance will the concept of mechanism exhaust the essence of the phenomenon, and adequately explain it; as for the other two phenomena, this model will explain only *certain* facets and aspects, or a certain fetishised form of them, or perhaps it will offer a first approximation and a potential way of conceptually grasping them. These phenomena are instances of a more complex reality whose adequate description and interpretation calls for structurally adequate logical categories (models).

It is important that contemporary philosophy know how to pick out the *real* central issues and the content of concepts introduced in the varied, unclear and frequently mystifying terminology of different philosophical schools and tendencies. It should examine whether classical concepts of materialist philosophy, e.g. totality, are not more suitable for conceptually grasping problems of contemporary science described in terms of structure and system. Both of these concepts might be implied in the concept of concrete totality.

From this perspective one might also criticize the inconsistencies and the biases of those philosophical tendencies which reflect in a certain way the spontaneous genesis of dialectics from twentieth century science (Lenin).

Such is the philosophy of the Swiss thinker Gonseth. Gonseth emphasizes the dialectical character of human cognition but his fear of metaphysics prevents him from satisfactorily establishing whether or not the objective reality that human thinking comes to know is itself dialectical. According to Gonseth, human cognition arrives at different horizons or images of reality but never reaches the 'ultimate' reality of things. If he meant that reality cannot be exhausted by human cognition, and that it is an absolute totality, whereas at every stage of its development mankind reaches only a certain relative totality, i.e. captures reality only to a certain degree, we could agree with Gonseth. Some of his formulations have, however, an explicitly relativistic character. Man's cognition has apparently nothing to do with reality itself but only with certain horizons or images of reality. These are historically variable but they never capture the fundamental, 'ultimate' structure of reality. Reality thus evaporates and man is left only with its image. Gonseth improperly confuses the ontological question and the gnoseological one, the question of objective truth and the dialectic of absolute and relative truth, as evident e.g. from the following clear formulation: 'The natural world is such, and we are such, that reality is not given to us in complete cognition [which is correct], in its essence [which is incorrect]'.[18] Cognition that is severed from nature, matter and objective reality cannot but fall into a degree of relativism, for it is never more than the cognition or expression of images or horizons of reality, and cannot formulate or recognize how objective reality itself comes to be known through these horizons or images.

The methodological principle for dialectically investigating objective reality is the standpoint of concrete totality. This implies that every phenomenon can be conceived as a moment of a whole. A social phenomenon is an historical fact to the extent to which it is studied as a moment of a certain whole, that is, to the extent to which it fulfils that *two-fold* role which makes it an historical fact in the first place: the role of defining itself and of defining the whole; of being both the producer and the product; of determining and being determined; of exposing while being decoded; of acquiring proper meaning while conveying the sense of something else. This interconnectedness and mediatedness of the parts and the whole also signifies that isolated facts are abstractions, artificially uprooted moments of a whole which become concrete and true only when set in the respective whole. Similarly, a whole whose moments have not been differentiated and determined is merely an abstract, empty whole.

The distinction between systematic-additive cognition and dialectical cognition is essentially the distinction between two different conceptions of reality. If reality were a sum of facts, then human cognition could amount only to abstract, systematic-analytic cognition of abstract parts of reality, whereas the whole of reality would remain unknowable. 'The object of scientific inquiry', says Hayek in his polemic with Marxism, 'is never the totality of all observable phenomena in a given time and space, but always only certain selected aspects of it . . . The human spirit can never encompass the 'whole' in the sense of all different aspects of the real situation'.[19]

Precisely because reality is a structured, evolving, and self-forming whole, the cognition of a fact or of a set of facts is the cognition of their place in the totality of this reality. In distinction from the summative-systematic cognition of rationalism and empiricism which starts from secure premises and proceeds systematically to array additional facts, dialectical thinking assumes that human cognition proceeds in a spiral movement in which *any* beginning is abstract and relative. If reality is a dialectical, structured whole, then concrete cognition of reality does not amount to systematically arraying facts with facts and findings with findings; rather, it is a process of *concretization* which proceeds from the whole to its parts and from the parts to the whole, from phenomena to the essence and from the essence to phenomena, from totality to contradictions and from contradictions to totality. It arrives at concreteness precisely in this spiral process of totalization in which all concepts move *with respect to one another*, and mutually illuminate one another. Neither does further progress of dialectical cognition leave individual concepts untouched; such cognition is not a summative systematization of concepts erected upon an immutable basis, constructed once and for all, but is rather a spiral process of interpenetration and mutual illumination of concepts, a process of dialectical, quantitative–qualitative, regressive–progressive totalization that transcends abstractness (one-sidedness and isolation). A dialectical conception of totality means that the parts not only internally interact and interconnect both among themselves and with the whole, but also that the whole cannot be petrified in an abstraction superior to the facts, because precisely in the interaction of its parts does the whole *form* itself as a whole.

Opinions as to whether concreteness as the cognition of all facts is knowable or not are based on the rationalist–empiricist idea that cognition proceeds by the analytic–summative method. This idea is in turn based on the atomist idea of reality as a sum of things, processes and facts. Dialectical thinking, by contrast, grasps and depicts reality as a whole that is not *only* a sum of relations, facts and processes, but is also the very *process of forming* *them**, their structure and their genesis. The *process of forming* the whole and of forming a unity, the unity of contradictions and its genesis, all belong to the dialectical whole. Heraclitus expressed the dialectical concept of reality in a great metaphor of the cosmos as a fire kindled and quenched according to rule, and he especially emphasized the *negativity* of reality: he described fire as 'need and satiety'.[20]

Three basic concepts of the whole, or totality, have appeared in the history of philosophical thinking, each based on a particular concept of reality and postulating corresponding epistemological principles:

(1) the *atomist-rationalist* conception, from Descartes to Wittgenstein, which holds reality to be a totality of simplest elements and facts;

(2) the *organicist and organicist-dynamic* conception which formalizes the whole and emphasizes the predominance and priority of the whole over its parts (Schelling, Spann);

(3) the *dialectical conception* (Heraclitus, Hegel, Marx) which grasps reality as a structured, evolving and self-forming whole.

The concept of totality has been attacked from two sides in the twentieth century. For empiricists, as for existentialists, the world has collapsed, it has ceased to be a totality and has turned into chaos. Organizing it is a matter for the subject. This transcendental subject or the subjective perspective, for which the totality of the world has collapsed and has been substituted by a scatter of subjective horizons, introduces order into the world's chaos.[21]

The subject who comes to know the world and for whom the world exists as the cosmos, divine order, or as totality, is *always* a social subject, and the activity of knowing the natural and the socio-human reality is the activity of a social subject. Severing society from nature goes hand in hand with not grasping that socio-human reality is *equally* a reality as nebulae, atoms or stars are, although it is not an *equal* reality. The suggestion will follow that the reality of nature is the only real one, and that human reality is less real than that of rocks, meteorites or suns; or that only one reality (the human one) can be comprehended, whereas the 'other' (the natural one) can at best be explained.

---

*Translating the concept of *Bildung* into Czech is as problematic as translating it into English. Kosík employed the word 'vytvářet', one specific form (the imperfective aspect) of the word 'tvořit', 'to create', 'to form'. Concepts related to Bildung have been rendered as *the process of forming, to form,* and *formative.*

According to materialism, social reality is known in its concreteness (totality) at the point when the *character* of social reality is exposed, when the pseudoconcrete is abolished and when social reality is known as the dialectical unity of the base and the superstructure, with man as its objective, socio-historical subject. Social reality is *not* known as a concrete totality as long as man is intuited primarily or exclusively as an object in the framework of totality, and as long as the primary importance of man as the *subject* of mankind's objective-historical praxis remains unrecognized. The concreteness, the totality of reality is thus not a matter of whether the facts are complete and whether horizons can change and shift; rather, it involves a *fundamental* question: *What is reality?* As for social reality, *this* question can be answered when reduced to a *different* one: How is social reality *formed?* This type of questioning, which establishes what social reality *is* by way of establishing how it is *formed*, contains a *revolutionary* concept of society and man.

Turning back to the question of the fact and its importance for the cognition of social reality, we have to emphasize (apart from the generally acknowledged position that every *fact* is comprehensible only in context and in a whole[22]) one other even more important and more fundamental point which is usually ignored: that *the very concept of fact is determined by the overall conception of social reality*. What an historical fact is, is only a partial question of the main one: What is social reality?

We agree with the Soviet historian I. Kon, that elementary facts have turned out to be something very complex, and that science which in the past used to deal with unique facts is now orienting itself more and more toward processes and relations. The relationship between facts and their generalizations is one of interconnection and interdependence; just as generalizations would be impossible without facts, there are no scientific facts that would not contain an element of generalization. An historical fact is in a sense not only the prerequisite for investigation *but* is also its result.[23] However, if facts and generalizations dialectically interpenetrate, if every fact carries elements of generalization and if every generalization is a generalization of facts, how is one to explain this *logical* mutuality? This logical relationship expresses the fact that a generalization is the *internal connection* of facts and that a fact itself mirrors a certain complex. The ontological essence of every fact reflects the whole reality, and the *objective* significance of a fact depends on how richly and how essentially it both encompasses and mirrors reality. This is *why* one fact can state more than another fact. This, too, is why it has more to state or less, according to the

method and the subjective approach of the scientist, i.e. according to how well the scientist questions the objective content and significance of his fact. Dividing facts by significance and importance follows not from subjective judgement but from the objective content of the facts themselves. Reality exists in a certain sense only as a sum of facts, as a hierarchized and differentiated totality of facts. Every cognitive process of social reality is a circular movement. Investigation both starts from the facts and comes back to them. Does something happen to the facts in the process of cognition? Cognition of historical reality is a process of theoretical appropriation, i.e. a critique, interpretation and evaluation of facts; an indispensable prerequisite of *objective* cognition is *the activity* of man, the scientist. This activity, which discloses the *objective* content and meaning of facts, is the scientific method. A scientific method is fruitful to the degree to which it manages to expose, interpret and substantiate the wealth of reality that is *objectively* contained in this or that particular fact. The indifference of certain methods and tendencies to facts is well known; it is an inability to see in facts anything important, i.e. their proper objective content and meaning.

Scientific method is a means for decoding facts. How did it ever happen that facts are not transparent but pose a problem whose sense science must first expose? A fact is coded reality. Naive consciousness finds facts opaque because of their perpetual *two-fold* role, discussed above. To see only one facet of facts, either their immediacy or their mediatedness, either their determinacy or their determining character, is to encode the code, i.e. to not grasp the fact as a code. In the eyes of his contemporaries, a politician appears as a *great* politician. After his death it turns out that he was merely an *average* politician and that his apparent greatness was an 'illusion of the times'. What is the historical fact? The *illusions* that had *influenced* and 'created' history, or the truth that came into the open only subsequently, and at the crucial time had not existed, had not happened as a reality? An historian is to deal with events as they really happened. Yet, what does this mean? Is real history the history of people's consciousness, the history of how people were aware of their contemporary scene and of events, or is it an history of how events really occurred and how they *had to be* reflected in people's consciousness? There is a double danger here: one can either recount history as it *should have* happened, i.e. infuse it with rationality and logic, or one can describe events uncritically, without evaluation, which of course amounts to abandoning a fundamental feature of scientific work, namely the distinction between the essential and the peripheral, which is the *objective* sense of facts. The existence of science is based on the *possibility* of this distinction. There would be no science without it.

Mystification and people's false consciousness of events, of the present and the past, is a *part* of history. The historian who would consider false consciousness to be a secondary and a haphazard phenomenon and would deny a place in history to it as to something false and untrue would in fact be distorting history. While Enlightenment eliminated false consciousness from history and depicted the history of false consciousness as one of errors that could have been avoided if only people had been more farsighted and rulers wiser, romantic ideology, on the contrary, considered false consciousness to be true, to be the only one that had any effect and impact, and was *therefore* the only historical reality.[24]

Hypostatizing the whole and favoring it over its parts (over facts) is one path that leads to a *false totality* instead of to a *concrete* one. If the whole process represented a reality which would be indeed *genuine* and *higher* than facts, then reality could exist independently of facts, independently in particular of facts that would contradict it. The formulation that hypostatizes the whole over the facts and treats it autonomously provides a theoretical substantiation for subjectivism which in turn ignores facts and violates them in the name of a 'higher reality'. The facticity of facts is not their reality but rather their fixed superficiality, one-sidedness and immobility. The *reality* of facts is opposed to their facticity not so much as a reality of a *different* order and independent of facts, but rather as an *internal* relation, as the dynamics and the contradictory character of the totality of facts. Emphasizing the whole process over facts, ascribing to tendencies a reality higher than to facts, and the *consequent* transformation of a tendency of facts into a tendency independent of facts, are all expressions of a hypostatized whole predominant over its parts, and thus of a *false* totality predominant over the concrete totality. If the process as a whole amounted to a reality higher than facts, rather than to the reality and lawfulness of facts *themselves*, it would become independent of facts and would lead an existence different from theirs. The whole would be separated from the facts and would exist independently of them.[25]

Materialist theory distinguishes between facts in two different contexts: in the context of reality where facts are set primordially and originally, and in the context of theory where they are arrayed secondarily and mediately, after having been torn out of the original context. But how can one discuss a context of reality where facts are originally and primordially, if the *only* way to know *this* context is through facts that have been *torn out* of it? Man cannot know the context of reality other than by extracting facts from it, isolating them and making them relatively autonomous. This is the basis of *all* cognition: dividing the one. All cognition is a dialectical oscillation (dialectical as opposed to metaphysical, for which both poles would be constant magnitudes and which would record their external, reflexive relations), and oscillation between facts and context (totality), an oscillation whose mediating active center is the method of investigation. Absolutizing the activity of the method (about this activity itself there is no doubt) begets the idealistic illusion that thinking generates the concrete, or that facts first acquire sense and significance *only* in man's head.

The fundamental question of materialist epistemology[26] concerns the relation of concrete and abstract totalities and the possibility of one changing into the other: how can the thought process of intellectually reproducing reality stay on the level of *concrete* totality, and not sink into an abstract totality? When reality is radically severed from facticity, it is hard to recognize *new* tendencies and contradictions in facts: because even before it investigates anything, *false* totality considers every fact to be predetermined by a once-and-for-all established and hypostatized evolutionary tendency. Despite its claims to a *higher* order of reality, this tendency will itself degenerate into an abstraction, i.e. into a reality of a *lower* order than is that of empirical facts, if it is conceived of not as an historical tendency of facts *themselves* but as one existing beyond, outside, above and independently of facts.

False totalization and synthetization show up in the method of the abstract principle which leaves aside the wealth of reality, i.e. its contradictory character and its multiple meanings, and deals only with facts that accord with this abstract principle. The totality to which this abstract principle might be promoted amounts to an *empty* totality which treats the wealth of reality as an irrational 'residue' beyond comprehension. The method of the 'abstract principle' distorts the *whole* picture of reality (of an historical event, of a work of art) and is *equally* insensitive to its *details*. It is aware of particulars, registers them, but does not understand them since it fails to grasp their significance. Instead of uncovering the objective sense of facts (details), it obfuscates it. It abolishes the wholeness of the investigated phenomenon by decomposing it into two autonomous parts: that which agrees with the principle and can be interpreted by it, and that which contradicts the principle and therefore remains in darkness (with no rational explanation or comprehension of it), as an unilluminated and unclarified 'residue' of the phenomenon.

The standpoint of concrete totality has nothing to do with the holistic, organicist, or the neo-romantic concepts of wholeness which hypostatize the whole over its parts and mythologize it.[27] Dialectics cannot grasp totality as a ready-made or formalized whole determining the parts because the *genesis* and *development* of totality are components of its very determination. From the methodological perspective, this calls for an examination of how totality *originates* and of the *internal sources of its development* and movement. Totality is not a ready-made whole, later filled with a content and with properties and relations of its parts; rather, totality *concretizes* itself *in the process of forming its whole as well as its content.* The genetic-dynamic character of totality is emphasized in the remarkable fragments of Marx's *Grundrisse*: 'While in the completed bourgeois system every economic relation presupposes every other in its bourgeois economic form, and everything posited is thus also a presupposition, this is the case with every organic system. This organic system itself, as a totality, has its presuppositions, and its *development* to its totality consists precisely in subordinating all elements of society to itself, or in creating out of it the organs which it still lacks. *This is historically how it becomes a totality.* The process of becoming this totality forms a moment of its process, of its development'.[28]

The genetic-dynamic conception of totality is a prerequisite for rationally grasping the genesis of a new quality. Prerequisites that originally had been historical conditions for the genesis of capital, appear after its emergence and constitution as results of capital's own self-realization and reproduction. They are no longer conditions of its *historical genesis* as much as results and conditions of its *historical existence*. Individual elements (such as money, value, exchange, labor power) that historically preceded the emergence of capitalism, that had existed independently of it and compared with capitalism had led an 'antediluvian' existence, are after the emergence of capital incorporated into the process of its reproduction and exist as *its* organic moments. Thus in the epoch of capitalism, capital turns into a structure of meanings that *determines* the internal content and the objective sense of its elements, a content and sense that in the pre-capitalist phase had been different. *The forming of a totality as a structure of meanings is thus also a process which forms the objective content and meaning of all its elements and parts as well.* This interconnection, as well as the profound difference of conditions of genesis (which are an independent, unique historical prerequisite) and of conditions of historical existence (which are historically produced and reproduced forms of existence), involve the dialectic of the logical and the historical: logical investigation indicates where historical investigation begins, and that in turn complements and presupposes the logical.

Insisting on the question of what is primary, whether totality or *contradictions*, or indeed dividing contemporary Marxists into two camps,[29] according to what they prefer, demonstrates an absolute lack of comprehension of materialist dialectics. The question is not whether to recognize the priority of totality over contradictions or vice versa, precisely because such a division strips both totality and contradictions of their dialectical character: *without contradictions, totality is empty and static; outside totality, contradictions are formal and arbitrary.* The dialectical relationship of contradictions and totality, of contradictions within totality and the totality of contradictions, of the concreteness of a totality formed by contradictions and the lawful character of contradictions within totality, all this is one of the distinctions that set apart the materialist and the structuralist conceptions of totality. Further: totality as a conceptual means of comprehending social phenomena is abstract as long as it is not stressed that this is a totality of the base and the superstructure, of their interrelation, mutual movement and development, with the base playing the determining role. And finally, even the totality of the base and the superstructure is abstract when it is not demonstrated that man is the *real historical subject* (i.e., of praxis), and that in the process of social production and reproduction he forms both the base and the superstructure, that he forms social reality as a totality of social relations, institutions and ideas, and that in this process of forming the objective social reality he also forms himself as an historical and social being with human senses and potentialities, realizing thereby the infinite process of 'humanizing man'.

Concrete totality, as the dialectical–materialist standpoint of the *cognition* of reality (we have several times emphasized its derivative character, compared with the ontological problem of reality), thus signifies a complex process with the following moments: destruction of the pseudoconcrete, i.e. of fetishist and fictitious objectivity of the phenomenon, and cognition of its real objectivity; further, the cognition of the phenomenon's historical character which in a peculiar way reveals the dialectic of the unique and of the generally human; and finally, the cognition of the objective content and meaning of the phenomenon, of its objective function and its historical place within the social whole. When cognition does not destroy the pseudoconcrete, when it does not expose the phenomenon's real historical objectivity under its fictitious objectivity, and when it consequently confuses the pseudoconcrete with the concrete, it becomes a captive of fetishist intuiting and results in a bad totality.[30] Social reality is then conceived of as a sum or a totality of autonomous structures influencing one another. The subject vanishes, or more precisely, the place of the real subject, i.e. of man as an objective—practical subject, is taken by a subject that has been mythologised, reified and fetishised: by the autonomous movement of structures. Materialistically conceived totality is formed by man's social production, while for structuralism, totality arises from the interaction of autonomous series of structures. In 'bad totality', social reality is intuited only in the form of the object, of ready-made results and facts, but not subjectively, as objective human praxis. The fruit of human activity is divorced from the activity itself. The dual movement from product to producer and from producer to product[31] in which the producer, creator, man, stands *above* his artifacts, is replaced in relativistic 'bad totality' by a simple or a complex movement of autonomous structures, i.e. of results and artifacts taken in isolation, through the objectivation of objective—intellectual human praxis. Consequently, in structuralist concepts 'society' enters into art only from without, as social determinism. It is not intrinsic to art, subjectively, as the social man who is its creator. Aside from idealism, the second basic feature of the structuralist conception of totality is sociologism.[32]

False totality appears in three basic forms:

(1) As *empty* totality which lacks reflection, the determination of individual moments, and analysis. Empty totality excludes reflection, i.e. the appropriation of reality as individual moments, and the activity of analytical reason.[33]

(2) As *abstract* totality which formalizes the whole as opposed to its parts and ascribes a 'higher reality' to hypostatized 'tendencies'. Totality thus conceived is without genesis and development, without the process of forming the whole, without structuration and destructuration. Totality is a *closed* whole.

(3) As *bad* totality, in which the real subject has been substituted by a mythologized subject.

Important concepts of materialist philosophy, such as false consciousness, reification, subject—object relationship, etc., lose their dialectical character when they are isolated, torn out of the materialist theory of history and severed from other concepts which together form a whole and an 'open system' that lends them real meaning. The category of totality also loses its dialectical character when it is conceived only 'horizontally', as the relation of parts and the whole, and when other of its organic features are neglected: such as its *'genetic—dynamic'* dimension (the forming of the whole and the unity of contradictions) and its *'vertical'* dimension (the dialectic of the phenomenon and the essence). The dialectic of the phenomenon and the essence was applied in Marx's analysis of simple capitalist commodity exchange. The most elementary and ordinary phenomenon of everyday life in a capitalist society – simple commodity exchange – in which people play the roles of simple buyers and sellers, shows under further investigation to be a superficial appearance that is determined and mediated by essential deep processes of the capitalist society – by the existence and the exploitation of wage labor. The freedom and equality of simple exchange is *developed and realized in the capitalist system of production as inequality and lack of freedom.* 'A worker who buys commodities for 3s. appears to the seller in the same function, in the same equality – in the form of 3s. – as the king who does the same. All distinction between them is extinguished'.[34]

The internal relation of the phenomenon and the essence, and the development of the contradictions of this relation, are dimensions which grasp the reality *concretely*, i.e. as a concrete totality. By contrast, hypostatizing reality's phenomenal aspects produces an abstract view and leads to apologetics.

NOTES

[1] The minds of people 'reflect always only the immediate *phenomenal forms* of relations, rather than their internal structure. If the latter were the case, of what use would science be?' (Marx's letter to Engels, 27 June 1867. Marx–Engels, *Werke*, Berlin, 1967ff., vol. 31, p. 313). '. . . all science would be superfluous if the outer appearance and the essence of things directly coincided.' (Marx, *Capital*, New York, 1967, vol. 3, p. 817.) 'For . . . the *phenomenal form,* . . . as contrasted with the *essential relation*, the same difference holds that holds with respect to all phenomena and their hidden substratum. The former appear directly and spontaneously as current modes of thought; the latter must first be discovered by science.' (Marx, *Capital*, vol. 1, p. 542; emph. Kosík)

[2] Certain philosophers (e.g. G. G. Granger, 'L'ancienne et la nouvelle économique', *Esprit*, 1956, p. 515) ascribe the 'method of abstraction' and of 'concept' exclusively to Hegel. In reality, this is the only path by which philosophy can arrive at the structure of the thing, i.e. to a *grasp* of it.

[3] Marx, Hegel, and Goethe were all advocates of this practical 'one-sidedness' opposed to the fictitious 'all-sidedness' of romanticists.

[4] Marx's *Capital* is methodologically constructed upon the distinction of false consciousness and the real grasping of things, and the main categories of conceptually grasping the reality under investigation are the following pairs:

phenomenon – essence
world of appearances – real world
external appearance of the phenomenon – law of the phenomenon
real existence – internal essential concealed kernel
visible movement – real internal movement
idea – concept
false consciousness – true consciousness
  doctrinaire systematization of ideas ('ideology') – theory and science.

[5] 'Marxism is an effort to detect behind the pseudo–immediacy of the reified economic world the social relations that formed it and that are concealed behind their own creation'. A. de Waelhens, *L'idée phénoménologique de l'intentionalité*, The Hague, 1959, p. 127f. The characterization offered by a non-Marxist author is a symptomatic testimony of philosophy in the twentieth century, for which the destruction of the pseudo-concrete and all manner of alienation has become a most pressing problem. Various philosophies differ in the *mode* of solving it, but the problematique itself is shared by both positivism (cf. Carnap's and Neurath's struggle against metaphysics, real or imagined), and phenomenology and existentialism. Characteristically, it took a Marxist philosopher, Tran-Duc-Thao, whose work was the first serious attempt to confront phenomenology and Marxism, to expose the authentic sense of Husserl's phenomenological method and its internal connection with philosophical problems of the twentieth century. Tran-Duc-Thao fittingly characterized the contradictory and paradoxical character of the phenomenological destruction of the pseudoconcrete: 'In the ordinary language, the world of appearances has arrogated the whole sense of the notion of reality . . . Appearances present themselves in the name of the real world and eliminating them took the form of bracketing the world . . . The authentic reality to which one was returning paradoxically took on the form of the irreality of pure consciousness'. Tran-Duc-Thao *Phénoménologie et materialisme dialectique*, Paris. 1951 pp. 223f. [Eng. trans. *Phenomenology and Dialectical Materialism*, D. Reidel, Dordrecht and Boston, forthcoming].

[6] Hegel has characterized reflexive thinking thus: 'Reflection is that form of mental activity which establishes the contradiction and which goes from the one to the other, but without effecting their combination and realizing their pervading unity'. Hegel, *Philosophy of Religion*, London, 1895, pp. 204f (adapted). See also Marx, *Grundrisse*, p. 88.

[7] Cf. Marx, 'Critique of Hegel's Doctrine of the State', in *Early Writings*, New York, 1975, p. 174 *et passim*.

[8] Positivism of the Viennese school played a positive role in destroying the pseudoconcrete, when it opposed surviving metaphysical conceptions by stating that matter is not something behind phenomena or the transcendence of phenomena, but that it is rather material objects and processes. Cf. Neurath, *Empirische Soziologie*, Vienna 1931, pp. 59–61 [Eng. trans. in *Empiricism and Sociology*, Vienna Circle Collection, Vol. 1, pp 358–64, D. Reidel, Dordrecht and Boston, 1973].

[9] This problematique will be further developed in chapters 'The Economic Factor' and 'Philosophy of Labor'.

[10] Polemics against dialectical materialism relentlessly impute to modern materialism the mechanical and metaphysical concept of matter of eighteenth-century theories. Why should *only* the spirit, and not matter, have the property of negativity? Sartre's thesis that materialism cannot be the philosophy of revolution (cf. his 'Materialism and Revolution', in his *Literary and Philosophical Essays*, New York, 1962, pp. 198–256) also stems from a metaphysical concept of matter, as *indirectly* acknowledged by Merleau-Ponty: 'Occasionally, the justified question is raised, how could materialism

possibly be dialectical (Sartre, 'Materialism and Revolution'), how could matter in the strict sense of the word contain the principle of productivity and of generating novelty, which is referred to as dialectics'. (*Temps modernes*, 1, p. 521.) All arguments concerning the acceptance or the rejection of the 'dialectics of nature' orbit around this question.

[11] The German word *entwickeln* is a translation of the Latin *explicatio* and means 'Unfolding, clear structuration of a whole that had been dark, muddled and mysterious'. (J. Hoffmeister, *Goethe und der deutsche Idealismus*, Leipzig, 1932, pp. 120f.) Both Goethe and Marx use the word in this sense.

[12] See Marx, *Capital*, vol. 1, p. 19.

[13] A detailed explication of the 'position of totality' as a methodological principle of Marx's philosophy is presented in Lukács' well-known *History and Class Consciousness*, Cambridge, Mass., 1971. L. Goldmann further developed Lukács' thought; see, e.g., *The Hidden God*, London, 1961.

[14] One classic example is Karl Mannheim and holistic structuralist theories that stem from his work.

[15] K. R. Popper, *Poverty of Historicism*, New York, 1964, p. 78.

[16] See Popper, op. cit.

[17] L. von Bertalanffy, 'General System Theory' in *General Systems*, 1, (1956), p. 1.

[18] F. Gonseth, 'Remarque sur l'idée de complementarité', *Dialectica*, 1948, p. 413.

[19] F. A. Hayek, *Scientisme et sciences sociales*, Paris, 1953, p. 79. [Counter Revolution in Science, Glencoe, 1952.]

[20] K. Freeman, ed., *Ancilla to the Pre-Socratic Philosophers*, Oxford, 1952, p. 65.

[21] Characteristically, the first major post-war philosophical clash between Marxism and idealism was over the problem of *totality*. There are clear practical considerations behind this theoretical argument: Can reality be changed in a revolutionary way? Can socio-human reality be changed in its foundations and as a whole, i.e. in its totality and totally, or are only partial changes practicable and real, with the whole being either an immutable entity or an elusive horizon? See the polemic between G. Lukács and K. Jaspers at 'Rencontres Internationales de Genève' of 1946, in J. Benda, ed., *L'Esprit Européen*, Neuchâtel, 1947.

The close connection between problems of totality and of revolution appears, appropriately modified, in Czech conditions as well: see K. Sabina's 1839 conception of totality as a revolutionary principle, in K. Kosík, *Česká radikální demokracie* [Czech Radical Democracy], Prague, 1958.

[22] See C. L. Becker, 'What are Historical Facts?', *Western Political Quarterly*, 8, 1955, no. 3, pp. 327–40.

[23] I. Kon, *Filosofskii idealism i krizis burzhoaznoi istoricheskoi mysli*, Moscow, 1959, p. 237.

[24] This is e.g. the error of H. Lévy-Bruhl in his essay 'Qu'est-ce que le fait historique?' *Revue de synthèse historique*, 42, 1926, pp. 53–59. I. Kon misinterprets Lévy-Bruhl's position, in his book mentioned above, and his polemic thus misfires.

[25] One can trace here the genesis of all objective idealistic mystifications. A valuable analysis of this problematique in Hegel is presented in E. Lask's *Fichte's Idealismus und Geschichte*, in Lask, *Gesammelte Schriften*, vol. 1, Tübingen, 1923, pp. 67f., 280, 338.

[26] For the time being we shall leave aside the question, how socio-human reality itself undergoes change and is transformed from a *concrete* to a *false* totality and vice versa.

[27] Schelling's great early thoughts about nature as a unity of product and productivity have not yet been sufficiently appreciated. Even at this stage, however, his thought demonstrates a strong tendency toward hypostatizing the whole, as evident from the following quote, dated 1799: 'Inasmuch as all parts of an organic whole carry and

support each other, this whole must have existed prior to its parts. The whole is not inferred from the parts, but the parts had to spring from the whole'. Schelling, *Werke*, Munich, 1927, vol. 2, p. 279.

[28] Marx, *Grundrisse*, p. 278 (emph. K. Kosík). [Penguin Books ed., 1973.]

[29] This opinion appeared at the international philosophical colloquium on dialectics in Royaumont, France, in September 1960. My paper 'Dialectique du concret' polemicised with this view.

[30] The term 'bad totality' was coined by Kurt Konrad who in his magnificent polemic against formalism discriminated between the concrete totality of materialism and the false bad totality of structuralism. See Kurt Konrad, *Svár obsahu a formy* [The Dispute of Content and Form], Prague, 1934.

[31] Cf. Leibniz: 'C'est par considération des ouvrages qu'on peut découvrir l'ouvrier'.

[32] This issue will be dealt with in detail in the chapter 'Historism and Historicism'.

[33] A critique of the economic concept of totality, for which all cats are black, was offered in Hegel's argument with Schelling, in his 'Introduction' to the *Phenomenology of the Mind*. Romanticists are obsessed with totality, but theirs is an empty totality because it lacks the fullness and determinacy of relations. Since the romanticist can absolutize the immediate, he can spare himself the journey from the particular to the general and arrives at everything — God, the Absolute, life — as by a shot of a gun. This is the main reason for the futility with which romanticists attempted to write a novel. The relationship of the vacuous totality of the romanticists and romanticist art is dealt with in B. von Arx, *Novelistisches Dasein*, Zürich, 1953, pp. 90, 96.

[34] Marx, *Grundrisse*, p. 246; cf. also p. 251.

CHAPTER II

ECONOMICS AND PHILOSOPHY

One wonders how appropriate is an investigation that reaches directly for the essence and leaves all the inessential behind as just excess baggage. Such investigation pretends to be something it is not. It claims to be scientific, yet it takes the most essential thing – the distinction between what is essential and what is peripheral – for granted and beyond investigating. It does not strive for the essential through a complex process of regressing and progressing which would at once cleave reality into the essential and the peripheral and substantiate such cleaving. Instead, it leaps over phenomenal appearances without ever investigating them and in so doing seeks to know both the essence and how to reach it. The directness of 'essential' thought skips the essential. Its chase after the essential ends in hunting down a thing without its essence, a mere abstraction or triviality.

Before an individual ever reads a textbook of political economy and learns about the scientifically formulated laws of economic phenomena, he *already* lives in an economic reality and understands it in his own way. Perhaps our investigation should then start by questioning the untutored individual? What promise might his answers hold, though? He might answer the question 'What is economics?' in words expressing his idea of it or regurgitating the answers of others. His answers will be mere echoes of those read or heard elsewhere. Similarly, his idea of economics will hardly be an original one, since its content will not measure up to reality. He who lives closest to economic reality and experiences it all his life does not necessarily have a correct idea of economics, i.e. of what he lives in. Important for the authenticity of our further reasoning is not how people answer the question about economics but rather what economics *is* to them, prior to any questioning and any contemplation. One always has a certain understanding of reality that precedes explication. Itself an elementary layer of consciousness, this pre-theoretical understanding is the basis for the *possibility* of the culture and the cultivation through which one ascends from a preliminary understanding to a conceptual cognition of reality. The belief that reality in its phenomenal appearance is a peripheral and negligible issue for philosophical cognition and for man leads to a fundamental error: ignoring the phenomenal appearance amounts to closing the door to the cognition of reality.

To investigate how economics exists for man is also to seek the most fundamental mode of this reality's givenness. Before economics becomes a topic for scientific considerations, explanations and interpretations, it already exists for man in a particular manifestation.

METAPHYSICS OF EVERYDAY LIFE

Care[1]

The primary and elementary mode in which economics exists for man is care. Man does not take care but care takes care of man. One is not careworn or carefree; rather, care *is* both in the careworn and in the carefree. Man may free himself of care but cannot set care aside. 'In life man belongs to care,' Herder has said. What then is care? To start with, care is not a psychological state or a negative frame of mind which would alternate with a different, positive one. Care is the subjectively transposed reality of man as an objective subject. Man is always already enmeshed in situations and relationships through his existence which is one of activity – though it may manifest itself as absolute passivity and abstention. Care is the entanglement of the individual in a network of relationships that confront him as the practical-utilitarian world. Therefore, objective relationships manifest themselves to the individual – in his 'praxis' rather than in his intuiting – as a world of procuring, of means, ends, projects, obstacles and successes. Care is the pure activity of the social individual in isolation. Reality cannot primarily and immediately manifest itself to this involved subject as a set of objective laws to which he is subjected; on the contrary, it appears as activity and interference, as a world which only the active involvement of the individual sets in motion and gives sense to. This world is *formed* through the involvement of the individual. Far from being merely a set of ideas, it is above all a certain kind of praxis in its most varied modifications.

Care is not the everyday consciousness of the struggling individual, one that he would shed during leisure. Care is the practical involvement of the individual in a tangle of social relations conceived from the position of his personal, individual, subjective involvement. These relations are not objectivised: they are not the subject-matter of science or of objective investigation, but are rather the sphere of individual involvement. Therefore the subject cannot intuit them as objective laws of processes and of phenomena; from the perspective of his subjectivity, he sees them as a world *related* to the subject, having meaning for this subject, and created by the subject. Since care is the entanglement of the individual in social relations seen from the perspective of the involved subject, it also amounts to a trans-subjective world seen by that subject. Care is the world in the subject. The individual is not only that which he considers himself or the world to be: he is also a part of the situations in which he plays an objective trans-individual role of which he may be quite unaware. In his subjectivity, man as care is outside himself, aiming at something else, transcending his subjectivity. Yet man is subjectivity not only in being outside himself and in transcending himself through it. Man's transcendence means that through his activity he is trans-subjective and trans-individual. His life-long care (*cura*) contains both the earthly element, directed at the material, and the element aspiring upward, to the divine;[1a] 'care' is ambiguous, and the question arises: Why this ambiguity? Is it a product and an artifact of Christian theological thought for which the ordeal of this world marks the only sure path to God? Is theology a mystified anthropology, or is anthropology a secularized theology? Theology can be secularized only because theological topics are in reality mystified problems of anthropology. Man's spanning of the earthly and the divine elements is a consequence of the dual nature of human praxis, which in its subjectively mystified form appears as the duality of 'care'.

The subject is determined by a system of objective relations, but acts as a concerned individual whose activity *forms* a network of relations. Care is:

(1) the entanglement of the social individual in a system of social relations on basis of his involvement and his utilitarian praxis;

(2) the activity of this individual which in the elementary form appears as caring and procuring;

(3) the subject of activity (of procuring and caring) which appears as lack of differentiation and anonymity.

Procuring is the phenomenal aspect of abstract labor. Labor has been divided up and depersonalized to the extent that in all its spheres — material, administrative, and intellectual — it appears as mere procuring and manipulation. To observe that the place occupied in German classical philosophy by the category of labor has been taken over in the twentieth century by mere procuring, and to view this metamorphosis as a process of decadence represented by the shift from Hegel's objective idealism to Heidegger's subjective idealism, is to highlight a certain *phenomenal* aspect of the historical process. The substitution of 'procuring' for labor does not reflect the qualities of a particular philosopher's thought or of philosophy as such; rather, it expresses in a certain way changes in the objective reality itself. The shift from 'labor' to 'procuring' reflects in a mystified fashion the process of intensified fetishization of human relations, a fetishization through which the human world reveals itself to the everyday consciousness (as fixed in a philosophical ideology) as a *ready-made* world of devices, implements and relations, a stage for the individual's social movement, for his initiative, employment, ubiquity, sweat, in one word – as procuring. The individual moves about in a *ready-made system of devices and implements*, procures them as they in turn procure him, and has long ago 'lost' any awareness of this world being a product of man. Procuring permeates his *entire* life. Work has been fragmented into a thousand independent operations, each of them with its *own* operator and executor, be it a production or a white-collar job. The manipulator faces not the work but an abstractly disintegrated segment of it which does not provide an overview of the work as a whole. The manipulator perceives the whole as a *ready-made* thing; of its genesis there exist only details, and these are in and of themselves irrational.

Procuring is praxis in its *phenomenally alienated form* which does not point to the *genesis* of the human world (the world of people and of human culture, of a culture that humanizes nature) but rather expresses the praxis of everyday manipulation, with man employed in a system of *ready-made* 'things', i.e., implements. In this system of implements, man himself becomes an object of manipulation. The praxis of manipulation (procuring) transforms people into manipulators and into objects of manipulation.

Procuring is manipulation (of things and of people). Its motions repeat daily, they have long ago become a habit and are performed mechanically. The reified character of praxis expressed in the term 'procuring' signifies that manipulation is not a matter of creating a work but of a man who, consumed by procuring, 'does not think' about the work. Procuring is man's practical behavior in a world that is ready-made and given; it amounts to attending and manipulating implements in a world, but in no way to the *process of forming* a human world. The philosophy that had offered a description of the world of care and procuring met with extraordinary acclaim because this particular world is the universal surface level of twentieth century reality. This world does not appear to man as the reality that *he* would have *formed* but as a ready-made and impenetrable world in which manipulation appears as involvement and activity. An individual manipulates the telephone, the automobile or the electric switch as something ordinary and unquestioned. It takes a break-down for him to discover that he lives in a world of *functioning* implements which constitute a mutually interlocking and interconnected system. A break-down indicates that 'implements' exist not in the singular but in the plural: that the telephone receiver is useless without the mouthpiece, the mouthpiece without the wiring, the wiring without electric current, current without the power station, the power station without coal (raw material) and machinery. A hammer or a sickle are not implements (apparatuses). Breaking a hammer is a perfectly transparent matter with which a single person can deal. A hammer is not an implement but a tool: it points not to a whole system of implements conditioning its own functioning but to the smallest circle of producers. In the patriarchal world of the plane, the hammer and the saw it is impossible to capture the problems of implements and apparatuses created by the modern industrial world of the twentieth century.[2]

Procuring as abstract human labor in its phenomenal form creates an equally abstract world of utility in which everything is transformed into a utilitarian instrument. In this world, things have no independent meaning and no objective being; they acquire meaning only insofar as they are manipulable. In practical manipulation (i.e. in procuring) things and people *are* implements, objects of manipulation, and acquire a meaning only in a system of general manipulability. The world discloses itself to the concerned individual as a system of meanings all of which point to all others, and the system as a whole points back to the subject for whom things have these meanings. This reflects, first, the complexity of modern civilization in which particularity has been transcended and its place taken by absolute universality. Second, behind the phenomenal form of the world of meanings (which when absolutized and separated from objective objectivity leads to idealism) there transpire the contours of the world of man's objective praxis and of its artifacts. In this world of meanings, the objective material praxis forms not only the meanings of things as the sense of things, but also the human senses which negotiate man's access to the objective meaning of things. The objective–practical and the sensory–practical world has dissolved in the perspective of care and has been transformed into a world of meanings outlined by human subjectivity. This is a *static* world in which manipulation, procuring and utilitarian calculation represent the movement of the concerned individual in a ready-made and fixed reality whose genesis is obscured. The *bond* of the individual with social reality is expressed and realized through care; but this reality discloses itself to concerned consciousness as a reified world of manipulation and procuring. Procuring as the universal reified image of human praxis is not *the process of producing and forming* an objective–practical human world, but is rather the manipulation of ready-made implements as of the total of civilization's resources and requirements. The world of human praxis is objective-human reality in its genesis, production and reproduction, whereas the world of procuring is one of ready-made implements and their manipulation. Since both the worker and the capitalist live in this twentieth century world of procuring, the philosophy of *this* world might appear to be more universal than the philosophy of human praxis. This fictitious universality results from its being a philosophy of *mystified* praxis, of praxis not as a human, transforming activity, but as the manipulation of things and people. Man as care is not merely 'thrown' into the world that is already there as a ready-made reality; rather, he moves about in this world — itself a creation of man — as in a complex of instruments he knows how to manipulate even without knowing their functioning and the truth of their being. In the process of procuring, man as care manipulates the telephone, the TV set, the elevator, the car and the subway, oblivious of the reality of technology and of the sense of these instruments.

Man as care is involved in social relations and *at the same time* has a certain relationship with nature and develops a certain idea of nature. Recognizing the human world as one of *utility* reveals an important truth: that this is a *social* world, in which nature appears as humanized nature, i.e. as the object and material base for industry. Nature is the laboratory and raw-material base for procuring, and man's relationship with it resembles that of a conqueror's relationship, a creator to his material. This, however, is only one of all *possible* relations, and the image of nature based on it exhausts neither the truth of nature nor the truth of man. 'Nature is sometimes reduced to being a workshop and to providing raw material for man's productive activity. This really is how nature appears to man — the producer. But the entirety of nature and its significance cannot be reduced to this role only. Reducing the relationship between man and nature to that of a producer and his raw material would infinitely impoverish human life. Such a reduction would indicate that the esthetic aspects of human life and of man's relation with the world have been uprooted — and more: the loss of nature as something created neither by man nor by anyone else, as something eternal and uncreated, would be coupled with the loss of the awareness that man is a part of a greater whole: compared with it, man becomes aware both of his smallness and of his greatness.'[3]

In care, the individual is always already in the future and turns the present into a means or a tool for the realization of projects. Care as the individual's practical involvement favors the future in a certain way, and turns it into the basic time dimension, in whose light he grasps and 'realizes' the present. The individual appraises the present and the past by the practical projects he lives for, by his plans, hopes, fears, expectations and goals. Since care is anticipation, it invalidates the present and fastens onto the future which has *not yet* happened. Man's time dimension, and his being as a being in time, are disclosed in care as a fetishised future and fetishised temporality: because it is ahead of the present, care considers the present not as the authentic existence, as 'closeness to being', but rather as a flight.[4] Care does not reveal the authentic character of human time. In and of itself, the future does not overcome romanticism or alienation. In a certain way it even amounts to an alienated escape from alienation, i.e. to fictitiously overcoming it. 'To live in the future', 'to anticipate' in a sense denies life: the individual as care lives not his present but his future, and since he neglects that which is and anticipates that which is not, his life occurs in nothingness, i.e. in inauthenticity, while he himself staggers between blind 'resoluteness' and resigned 'waiting'. Montaigne knew this form of alienation well.[5]

The Everyday and History

Every mode of human existence or being-in-the-world has its everyday. The Middle Ages had its everyday which was segmented among different classes, estates and corporations. Though the everyday of the serf differed from those of the monk, the wandering knight or the feudal lord, they all shared a common denomination, one single basis determining the tempo, rhythm, and organization of life – the feudal society. Industry and capitalism introduced not only new tools of production, new classes and political institutions but also a new manner of the everyday, one essentially different from that of previous epochs.

What is the everyday? The everyday is not privacy, as opposed to public life. Nor is it so-called profane life as opposed to an exalted official world: both the scribe and the emperor live in the everyday. Entire generations, millions of people have lived and still live the everyday of their lives as though it were a *natural* atmosphere, and they never pause to question its sense. What is the sense of questioning the sense of the everyday? Might such questioning perhaps suggest an approach that would expose the essence of the everyday? At what point does the everyday become problematic and what sense does this uncover? The everyday is above all the *organizing* of people's individual lives into every day: the replicability of their life functions is fixed in the replicability of every day, in the time schedule for every day. The everyday is the organizing of time and the rhythm which govern the unfolding of individual life histories. The everyday has its experience and wisdom, its sophistication, its forecasting. It has its replicability but also its special occasions, its routine but also its festivity. The everyday is thus not meant as a contrast to the unusual, the festive, the special, or to History: hypostatizing the everyday as a routine over History, as the exceptional, is itself the *result* of a certain mystification.

In the everyday, the activity and way of life are *transformed* into an instinctive, subconscious, unconscious and unreflected *mechanism* of acting and living: things, people, movements, tasks, environment, the world – they are not perceived in their originality and authenticity, they are not tested and discovered but they *simply are there*, and are accepted as inventory, as components of a *known* world. The everyday appears as the night of indifference, of the mechanical and the instinctive, i.e. as the world of familiarity. At the same time, the everyday is a world whose dimensions and potentialities an individual can control and calculate with his abilities and resources. In the everyday, everything is 'at hand' and an individual can realize his intentions. This is *why* it is a world of confidence, familiarity, and routine actions. Death, sickness, births, successes and failures are all accountable events of everyday life. In the everyday, the individual develops relations on basis of *his own* experience, *his own* possibilities, *his own* activity, and therefore considers the everyday reality to be his own world. Beyond the limits of this world of confidence, familiarity, immediate experience and replicability which the individual can count on and control, there begins another world, the very opposite to the everyday. The collision of these two worlds reveals the truth of each of them. The everyday becomes problematic and reveals itself as the everyday when it is disrupted. It is not disrupted by unexpected events or by negative phenomena: the exceptional and the festive on the level of the everyday are an integral part of it. Inasmuch as the everyday represents the organizing of millions of people's lives into a regular and replicable rhythm of work, action and life, it is disrupted only when millions of people are jolted out of this rhythm. War disrupts the everyday. It forcefully drags millions of people out of their environment, tears them away from their work, drives them out of their familiar world. Although war 'lives' on the horizon, in the memory and in the experience of everyday living, it is beyond the everyday.

War is History. In the collision of war (of History) with the everyday, the latter is overpowered: for millions, the customary rhythm of life is over. This collision of the everyday and History (war), in which one (particular) everyday has been disrupted and no other habitual, mechanical and instinctive rhythm of acting and living has yet been established, reveals both the character of the everyday and that of History, and their relationship.

Folk wisdom has it that one will even get used to the scaffold. That is, even in the most extraordinary, least natural and least human of environments, people develop a *rhythm* of life. Concentration camps had their everyday, and indeed even the person on death row has his. Two kinds of replicability and substitution operate in the everyday. Every day of the everyday can be substituted for another corresponding day, the everyday makes this Thursday indistinguishable from last Thursday or from last year's Thursday. It merges with other Thursdays and it would be preserved, i.e. it would differ and emerge in memory, only if there were something special and exceptional to it. At the same time, any subject of a given everyday can be substituted for any other subject: subjects of the everyday are interchangeable. They are best described and branded with a number and a stamp.

The clash of the everyday with History results in an upheaval. History (war) disrupts the everyday, but the everyday overpowers History – for *everything* has its everyday. In this clash, the separation of the everyday from history, a separation which is the starting and permanent vantage point of everyday consciousness, *proves* in practice to be a mystification. The everyday and history interpenetrate. Intertwined, their supposed or apparent character changes: the everyday no longer is that for which routine consciousness takes it, in the same way as History is not that as what it appears to routine consciousness. Naïve consciousness considers the everyday to be a natural atmosphere or a familiar reality, whereas History appears as a transcendental reality occurring behind its back and bursting into the everyday in form of a catastrophe into which an individual is thrown as 'fatally' as cattle are driven to the slaughterhouse. The *cleavage* of life between the everyday and History exists for this consciousness as fate. While the everyday appears as confidence, familiarity, proximity, as 'home', History appears as the derailment, the disruption of the everyday, as the exceptional and the strange. This cleavage simultaneously splits reality into the *historicity* of History and the *ahistoricity* of the everyday. History changes, the everyday remains. The everyday is the pedestal and the raw material of History. It supports and nourishes History but is itself devoid of history and outside of history. What are the circumstances of the everyday which transform it into the 'religion of the workaday', of acquiring the form of eternal and immutable conditions of human life? How did the everyday which is a product of history and a reservoir of historicity end up *severed* from History and considered the antinomy of history, i.e. of change and of events? The everyday is a *phenomenal* world which *reveals* reality in a certain way even as it *conceals* it.[6]

In a certain way, the everyday *reveals* the truth about reality, for reality outside the everyday world would amount to transcendental non-reality, i.e. to a formation without power or effectiveness: but in a way it also conceals it. Reality is contained in the everyday not immediately and in its totality but mediately and only in some aspects. An analysis of the everyday allows for reality to be grasped and described only to a *certain extent*. Beyond the limits of its 'potentialities' it falsifies reality. In this sense one grasps the everyday from reality, rather than vice versa.[7]

The method of the 'philosophy of care' is at once mystifying and demystifying in that it presents the everyday in a *particular* reality as though it were the everyday as such. It does not distinguish between the everyday and the 'religion' of the workaday, i.e. the alienated everyday. This method takes the everyday to be inauthentic historicity, and the transition to authenticity to be a rejection of the everyday.

If the everyday is the *phenomenal* 'layer' of reality, then the *reified* everyday is overcome not in a leap from the everyday to authenticity but in practically abolishing both the fetishism of the everyday and that of History, that is, in practically destroying reified reality both in its phenomenal appearance and in its real essence. We have demonstrated that radically separating the everyday from variability and historicity on the one hand leads to a *mystification of history* which then appears as the Emperor on horseback and as History, and on the other hand leads to *emptying the everyday*, to banality and to the 'religion of the workaday'. Divorced from history, the everyday becomes emptied to the point of being absurdly immutable. Divorced from the everyday, history turns into an absurdly *powerless* giant which bursts into the everyday as a catastrophe but which nevertheless cannot change it, i.e. cannot eliminate its banality or fill it with content. The plebeian naturalism of the nineteenth century believed that the importance of historical events lies not in how and why they developed but in how they influenced the 'masses'. But a mere projection of 'grand history' into the lives of ordinary people does not eliminate the idealistic view of history. It even strengthens it in a sense. From the point of view of official heroes, only the so-called exalted world, the world of grand deeds and of historical events which overshadow the emptiness of everyday life, rightfully belongs into history. Conversely, the naturalist concept negates this exalted world and focuses on a scatter of daily events, on mere records and documentary snapshots of ordinary life. This approach, however, deprives the everyday of its historical dimension as much as the idealistic approach does. The everyday is taken as eternal, in principle immutable, and thus compatible with any epoch in history.

The everyday appears as the anonymity and tyranny of the impersonal power which dictates every individual's behavior, thoughts, taste and even his protest against banality. The anonymity of the everyday, expressed in the subject of this anonymity, that is in *the someone/no-one*, has its counterpart in the anonymity of historical actors described as 'history makers'. Historical events consequently appear as the work of no-one and thus of all, as the result of anonymity shared both by the everyday and by History.

What does one mean by saying that the first and foremost subject of the individual is anonymity, that man understands himself and the world above all on basis of care and of procuring, on basis of the world of manipulation in which he is submerged? What does one mean by saying that 'Man *ist* das, was man betreibt'? What does it mean, that an individual is first immersed in the anonymity and facelessness of the someone/no-one which acts *in him*, thinks *in him*, protests *within him* on *his* behalf and on behalf of the *I*? Through his very existence, man is not only a social being which is already enmeshed in a network of social relations. He is also *acting, thinking* and *feeling* as a social *subject* even before he is or indeed could be aware of this reality. Routine consciousness (the 'religion') of the everyday takes human existence for a manipulable object and treats and interprets it accordingly. Since man identifies with his environment, with what is at hand, what he manipulates and what is ontically closest to him, his own existence and understanding of it turn into something distant and unfamiliar. Familiarity is an obstacle to knowledge. Man can figure out his immediate world of procuring and manipulation but cannot 'figure out' himself because he disappears in and merges with the manipulable world. The mystifying-demystifying 'philosophy of care' describes and postulates this reality but cannot explain it. Why does man first of all *disappear* in the 'external' world and interprets himself from it? Man is primordially what his world is. This derivative existence determines his consciousness and prescribes the way in which he is to interpret his own existence. The subject of an individual is first of all a *derivative* subject, both in terms of false individuality (the false

I) and false collectivity (the fetishised we). The materialist thesis which states that man is an ensemble of social conditions but neglects to mention who is the *subject* of these 'conditions'[8] leaves it to the 'interpretation' to fill in the blank either with a real or with a mystical subject, with the mystified I or the mystified we. Both transform the real individual into a tool and a mask.

The subject—object relationship in human existence is not identical with the relationship of the internal and the external, or with that of the isolated pre- or non-social subject and the social entity. The subject is already constitutively permeated with an objectivity which is the objectification of human praxis. An individual might be submerged in objectivity, in the world of manipulation and procuring, so completely that his subject disappears in it and objectivity itself stands out as the real, though mystified, subject. Man might disappear in the 'external' world because his is the existence of an objective subject which exists only by producing a subjective—objective historical world. Modern philosophy discovered the great truth that man is not born into conditions 'proper' but is always 'thrown'[9] into a world. He has to check *for himself* its authenticity or inauthenticity: in struggle, 'practical life', in the process of his own life history, in the course of appropriating and changing, of producing and reproducing reality.

In the course of the practical—spiritual evolution of the individual and of mankind, the undifferentiated and omnipotent rule of anonymity eventually collapses. In the course of ontogenesis and phylogenesis, its undifferentiated character diversifies into human and general human features on the one hand, the appropriation of which transforms an individual into a human individual, and into particular, non-human, historically transient features on the other hand, of which an individual has to free himself, if he is to work his way toward authenticity. In this sense, man's evolution progresses as a practical process of *separating* the human and the non-human, the authentic and the inauthentic.

We have characterized the everyday as a world with a regular rhythm in which man moves about following mechanical instincts, and with a feeling of familiarity. Reflection over the sense of the everyday leads to the absurd consciousness that there is no sense to it. 'What a bore to put on a shirt in the morning. Then the breeches over it. To crawl into bed at night and out again in the morning. To keep setting one foot in front of the other *with no prospect of it ever changing.* It's very sad. And to think that millions have done it before us and millions will do it again ...'[10] What is essential, however, is not the consciousness of the absurdity of the everyday, but the question of when does one come to reflect upon it. One questions the sense of the everyday with its automatism and immutability not because *it itself* would have become a problem. Rather, its problematization reflects a problematization of reality: primordially, one seeks not the sense of the everyday but the sense of reality. The feeling of absurdity is evoked not by reflection about the automatism of the everyday. Rather, reflection about the everyday is a *consequence* of the absurdity that historical reality has forced upon the individual (Danton).

Man *can* be man only if he can perform various life functions automatically. The less these activities impinge upon his consciousness and reflection, the better suited they are and the better service they render. The more complicated man's life, the more numerous are the relations he enters into; and the more functions he performs, the more extensive is the *necessary* sphere of automated human functions, customs, procedures. The process of automating and mechanizing the everyday of human life is an *historical* process. The boundary between the possible and necessary sphere of automation, on the one hand, and the sphere which in the best human interest cannot be automated, on the other hand, is consequently one that shifts in the course of history. With an increasingly complex civilization, man has to subject ever more extensive spheres of his activity to automation, in order to maintain enough space and time for genuine human problems.[11] The impossibility of automating certain life functions can be an obstacle to human life itself.

Inasmuch as the shift from the inauthentic to the authentic is an historical process which is realized both by mankind (a class, a society) and by the individual, an analysis of its concrete forms has to cover *both* of these processes. A forced reduction of one process to the other or their identification will transpire in the sterility and triviality of answers that philosophy might offer to the problems they pose.

The pseudoconcrete of the alienated everyday world is destroyed through *estrangement*, through *existential modification*, and through *revolutionary transformation*. Though this list does have an hierarchical aspect to it, every form of destruction maintains its relative independence, and to that extent cannot be substituted by another form.

The world of everyday familiarity is not a known and a recognized one. In order to *present* it in its reality, it has to be ripped out of fetishised intimacy and exposed in alienated brutality. Experiencing the workaday life naively and uncritically, as though it were the natural human environment, shares a substantial common trait with philosophical nihilism: in both, a particular historical form of the everyday is considered the natural and immutable basis for all human coexistence. In one instance, the alienation of the everyday is reflected in consciousness as an uncritical attitude, in the other as a feeling of absurdity. To behold the truth of the alienated everyday, one has to maintain a certain distance from it. To do away with its familiarity, one has to 'force' it. What is the kind of society and what is the kind of world whose people have to 'turn into' lice, dogs and apes in order for their real image to be represented adequately? In what 'forced' metaphors and parables must one *present* man and his world, to make people *see* their own faces and *recognize* their own world? One of the main principles of modern art, poetry and drama, of painting and film-making is, we feel, the 'forcing' of the everyday, the destruction of the pseudo-concrete.[12]

Presenting the truth about human reality is rightly felt to be something other than this reality itself, and it is therefore insufficient. It is not enough for the truth of reality to be *presented* to man; man has to *perform* this truth. Man wants to *live* in authenticity and to *realize* authenticity. An individual cannot by himself effect a revolutionary change in conditions and eradicate evil. Does this imply that as an individual, man has no immediate relationship to authenticity? Can he live an authentic life in a world that is inauthentic? Can he be free in an unfree world? Does there exist *one single* trans-personal and trans-individual authenticity, or is there a permanent choice, accessible to anyone and to all? In the existential modification, the subject of the individual awakens to his own possibilities and elects them. He changes not *the world, but his attitude toward it.* The existential modification is not a revolutionary transformation of the world but *the drama of an individual in the world.* In the existential modification, the individual liberates himself from the inauthentic existence and chooses an authentic one *among others*, by considering the everyday *sub specie mortis.* In that way he invalidates the everyday with all its alienation and rises above it, but at the same time he negates the *sense* of his own activity. Choosing authenticity *sub specie mortis* leads to aristocratic romantic stoicism (under the sign of death I live authentically, on the throne or in chains) or is realized as choosing death. *This* form of existential modification is, however, not the only way, or even the most frequent or the most adequate way for an individual's authentic realization. It, too, is only an historical choice with a quite precise social and class content.

METAPHYSICS OF SCIENCE AND REASON

Homo oeconomicus

Man as care is the pure subjectivity in which the whole world is submerged. In this chapter we shall trace the transition to the other extreme, to the subject who objectifies himself. In order to understand who he is, the subject becomes objectual (*objektální*)*. The subject is no longer mere involvement and activity that forms the world: now he becomes integrated in a transindividual lawlike whole as one of its components. However, this incorporation transforms the subject. The subject abstracts from his subjectivity and becomes an object and an element of the system. Man becomes a unit determined by its function in a lawlike system. He seeks to comprehend himself by abstracting from his subjectivity, and turns into an objectual being. The purely intellectual process of science transforms man into an abstract unit integrated in a scientifically analysable and mathematically describable system. This reflects the *real* metamorphosis of man performed by capitalism. Only under capitalism did economics develop as a *science.* Antiquity and the Middle Ages knew an economy, and a few scattered facts of economics, but not economics as a science.

The foremost question of modern science is, 'What is reality and how is it cognizable?' Galileo answered: All is real that can be described mathematically. To create a *science* of economics which would express the laws of economic phenomena, it was necessary to establish the turning point at which the individual becomes the general, the arbitrary the lawlike. The inception of political economy as a *science* fell in a period when the individual, the arbitrary and the random acquired the form of the necessary and the lawlike, when the totality of social movement arose 'from the conscious will and particular purposes of individuals', when it became *independent* of these purposes, and when 'the social relations of individuals to one another [appeared] as a power over the individuals that [had] become autonomous, whether conceived as a natural force, as chance or in whatever other form'.[13] Science (political economy) takes this emancipation of social movement as something primary, given and irreducible, and posits the task of describing the *laws* of this movement. The science of economic phenomena tacitly and unconsciously presupposes the idea of a *system*, i.e. of a certain differentiated whole whose laws can be traced and defined *just as* in the physical world. Thus the 'new science' is not presuppositionless; it

---

**objektální* is a Czech neologism of Kosík. The German translation renders it as *objekthaft.* –Tr.

is based on certain presuppositions, but ignores their significance and their *historical* character. Rightly or wrongly, the Physiocrats identified economics, conceived in its elementary *scientific* form with the bourgeois form of production. This was in turn studied in terms of the 'material laws' that arise from the character of production and are independent of will, politics, etc.[14] A theory of society as a system emerges only when society itself has become a system, when it has not only been sufficiently differentiated, but when this differentiation has led to multilateral *dependence*, and this dependence has itself become *independent* – i.e., when society is constituted as a differentiated whole. Capitalism is the first system in *this* sense of the word. Only on the basis of a reality grasped and comprehended in this way, in form of a natural order, i.e. only on the basis of economics as *a system of laws* that man studies, will one pose a secondary question, concerning man's *relation* to this system. *Homo oeconomicus is based on the idea of a system.* Homo oeconomicus is man as a component of a system, as a functioning element of a system, who as such must be equipped with essential features indispensable for running the system. The suggestion that the science of economic phenomena is based on psychology and that the laws of economics are just an elaboration, refining and objectivation of psychology[15] uncritically accepts the phenomenal form of reality as though it were reality itself. Classical science equipped the 'economic man' with several basic characteristics, including such fundamental ones as rational behavior and egoism. If the 'homo oeconomicus' of classical science is an abstraction, it is a reasonable abstraction: not only in the sense of *verständig* but especially in the sense of *vernünftig*. Its 'abstractness' is determined by the system, and only outside the system does homo oeconomicus become an abstraction devoid of content. *The system (economics as a system) and homo oeconomicus are inseparable magnitudes.* Helvetius' theory of interest and Ricardo's economic theory are based on a common foundation whose hidden character had led to many misunderstandings. Take for example the idea that the psychology of egoism (interest) – the laws of economics being definitions of a force called egoism – is directly analogous to a physical mechanism. Egoism can be considered the mainspring of human activity only in the framework of a system which takes it for granted that pursuing one's private interests will create general welfare. What is this 'general welfare' that appears as the *result*? It is the presupposition and the ideological *premise* that capitalism is the best system possible.

Interaction of as few as two people forms a system. More precisely, the interaction of two people is an elementary model of a social system. Mandeville's vain young lady and the crafty mercer, Diderot's Jacques Fataliste[15a] and his master, Hegel's master and slave, all represent certain concrete models of human relations, presented as a system. A system is more than the sum total of participants because people and their relations form something new, something transindividual, in a system and as a system. This is particularly conspicuous in Mandeville whose people are of a certain kind only inasmuch as they act; but they can act only in the framework of a particular system of relations which in turn presupposes, requires and shapes *particular people.*[16]

What kind of man, and with what psychological endowment, must the system form in order for it to function? Even if it does 'form' people with an instinct for earning and an instinct for saving, with rationalized behavior directed at maximum effect (utility, profit, etc.), it still does not follow that people are identical with these abstractions. Rather, it means that these basic characteristics are *sufficient* for the system to function. *Not theory, but reality itself reduces man to an abstraction. Economics is a system and a set of laws governing relations in which man is constantly being transformed into the 'economic man'.* Entering the realm of economics, man is *transformed.* The moment he enters into economic relations, he is drawn,— irrespective of his will and consciousness — into situations and lawlike relations in which he *functions* as the homo oeconomicus, in which he exists and realizes himself only to the extent to which he fulfills the role of the economic man. Thus economics is a sphere of life that has the tendency to transform man into the economic man and that draws him into an objective mechanism which subjugates and adapts him. Man is active in the economy only insofar as the economy is active, i.e. insofar as it *transforms man* into a certain abstraction, insofar as it absolutizes, exaggerates and emphasizes certain features while ignoring other, random ones which are unnecessary in the context of the economic system. This reveals how nonsensical are such contemplations that would divorce the 'economic man' from capitalism as a system. Homo oeconomicus is a fiction only when considered as an entity independent of the capitalist system.[17] As an element of the system, though, homo oeconomicus is a reality. *Thus classical economics begins not with the 'economic man' but with the system, and for the purposes of this system it posits the 'economic man' as a well-defined element of its construction and functioning.* Man is not defined in and of himself but with respect to the system. The primary question is not, 'What is man?' but rather, 'How does man have to be equipped for the system of economic relations to be set in motion and for it to function as a mechanism?' The concept of a system is a fundamental groundplan of science. Certain laws are exposed on its basis behind the apparent chaos of empirical phenomena. Before studying the empirical and factual nature of phenomena, there already exists the idea of a system, of an intelligible principle permitting the study of these phenomena. *Innumerable chaotic individual acts, seemingly arbitrary and random, are reduced to and interpreted as instances of a characteristic and typical movement.*[18] The introduction and application of the concept of a system is linked (1) with a certain scheme or model, an explicative principle of social phenomena, and (2) with quantification and mathematical methods, i.e. with the possibility of formulating economic laws in mathematical terms. It was in principle possible to introduce mathematics into economics because science takes economic phenomena to be a system of repetitive regularities and laws.

Classical economics presupposed a key turning point at which the subjective becomes the objective, and took it as a starting point without investigating it further. Questions of how this turning point might be possible and what exactly happens in it were not entertained. This unconcern contains a potential for mystification, and assorted protests against the 'reification' of man in classical political economy have been based precisely on this 'unconcern.' For classical economics, man exists exclusively as a part of the system, and *studies even himself* only by looking at himself as a part of the system. The ideal of scientific cognition of man consists in abstracting to the utmost from his subjectivity, from random features and idiosyncracies, of turning man into a 'physical magnitude' that can be constructed, described and eventually even formulated mathematically, as any other magnitude of classical mechanics.

The transition from man as 'care' to the 'economic man' is not merely a shift in perspective. The problem is not that in the first case man is intuited as subjectivity which knows nothing of the objectivity of social context, while in the second case that same man is investigated in an objective transindividual context. The main problem is elsewhere. With what appears as a shift in view or in perspective, the very subject-matter of the investigation changes, and *objective reality turns into an objectual reality*, a reality of objects. *Physis* turns into physics, nature is reduced to mere *natura naturata.* In what appears as a shift in *perspective*, man is transformed into an object and is investigated as though he were on the same level as any other thing or object. The human world turns into a physical world and the science of man turns into the science of man—object, i.e. into social physics.[19] A mere shift in perspective, intended to reveal certain aspects of reality, actually *forms a reality that is altogether different*, or rather, substitutes one thing for another while being oblivious of this substitution. The substitution involves more than the methodological approach to reality: *through the methodological approach, reality itself is changed. Methodology is ontologized.*[20] Vulgar economics is the ideology of an objectual world. It does not investigate its internal relations and laws but systematizes the *ideas* that agents of this objectual world, i.e. people reduced to objects, harbor about themselves, about the world and economics. Classical economics also deals with an objectual reality but rather than systematizing agents' ideas about this reified world, it searches for its internal laws. But if reification — the world of things and of reified human relations — is reality, and if science investigates it, describes it and searches for its internal laws, then what makes science itself fall for illusions and reification? This happens because science views this objectual world not merely as a particular form and as an historically transient period of human reality but *describes it* instead as *natural human reality.*

What appeared as a mere shift in perspective was in fact a substitution of realities: an objectual reality was *substituted* for an objective one.[21] *Social reality was conceived in terms of nature in its physical sense, and economic science in terms of social physics. Objective reality was therefore transformed into an objectual one, into a world of objects.*

The reality which classical economics describes by way of *its own method* is not an objective one. Classical economics does not describe the human world in its alienated form, nor does it demonstrate how socio-historical relations of people are masked by the relations and movement of things. Instead, *it describes this reified world and its laws as though it were the real human world, for this is the only human world of which classical economics is aware.*

Man becomes a reality only by becoming an element of the system. Outside the system he is unreal. He is real only to the extent to which he is reduced to a function of the system and to which the requirements of the system define him as homo oeconomicus. He is real only to the extent to which he cultivates those abilities, talents and inclinations that the system requires for its own operation. Other talents and capacities which are not indispensable for the system are superfluous and unreal. They are unreal in the true and original sense of the word. They cannot be actualized and realized, they cannot become the *real* activity of man, or transform into a real world for man to live in. They amount to an unreal world of privacy, irrelevance, of the romantic.

Romantic apologists have reproached Smith, saying that in his system, people are 'torn out of all natural and moral bonds, their relations are completely contractual, revocable, and assessable in terms of money. All that takes place among them is the market. They are so distilled a people that they hardly harbor any real drive for pleasure: only the drives for earning and saving move the economy'.[22] But posing the question in this way is foreign to classical economists as well as to Marx. It is a romantic reaction to the reality of capitalism. Classical economics sees the question thus: What necessary feature must man have for the capitalist system to function? By contrast, the romantic concept of a secondary system — which defines man from the system and reduces him to the system's requirements, leaving no place for the whole man to assert himself, since only *some* of his potentialities and functions can be realized in different spheres[23] — is a superficial, *degenerated* and romantic paraphrase of the classical theory. The fullness in whose name romantic apologists protest the abstract and distilled character of the 'economic man' is the fullness of a patriarchal man with undeveloped potential. Or does perhaps the free modern man see as his ideal a fullness that binds the individual from cradle to grave with a *single* organization in which he can develop his *limited* abilities? Is it not a great advantage of modern times that man can move about freely in *many* worlds and can (with certain historical and class limitations) transfer from one to another, that he is bound only by certain functions, and only for a limited time, to the 'organism' (i.e. to economics as a necessity of life), which is *precisely* how he cultivates his abilities? Is it not a manifestation of man's progress through history that he has the *capacity* to live simultaneously in several worlds, that he can perceive and experience different worlds? The fullness of modern man is of a different kind than that of the romanticized patriarchal man and it is found elsewhere. The fullness of earlier eras was in constraints on form and shape while the fullness of modern man is in the unity of diversities and contradictions. The very ability to act and live in more than one world is progress, when compared with guild constraints and constrained fullness. Romantic disparagement for systems and for abstraction forgets that the problem of man, of his freedom and his concreteness, is *always* one of his relation to the system. Man always exists in a system, and being one of its components he is reduced to certain aspects (functions) and to certain (one-sided and reified) forms of existence. At the same time, he is always more than a system, and *as man* he cannot be reduced to one. The existence of the concrete man spans the distance between his irreducibility to a system and the possiblity to transcend it, and his actual location and practical functioning in a particular system (of historical circumstances and relations).

Reason, Rationalization, Irrationality

The recurrent observation (Marx, Weber, Georg Lukács, C. Wright Mills) that the rationalization of modern capitalist society goes hand in hand with the loss of reason, and that with advancing rationalization irrationality spreads as well, correctly pinpoints an important symptom of our times. Yet, is it justified to juxtapose the reign of rationalization and irrationality against the 'independent reason of the Cartesian man'?[24] We shall see in the following that the independent reason of the Cartesian man is the product of rationalization and of irrationality. To juxtapose the consequence against the cause amounts to not beholding the essence of the problem. The question of how rationalization is transformed into a force that *excludes* reason, of how rationalization *begets* irrationality, can be systematically studied only by penetrating to the starting point of this inversion, i.e. by an historical analysis of reason.

Cartesian reason is the reason of a liberated isolated individual who finds in his own consciousness the only certainty of himself and of the world. This reason not only buttresses contemporary *science*, the science of rationalist reason, but also permeates contemporary *reality*, complete with its rationalization and irrationality. In its consequences and in its realization, 'independent reason' turns out to be dependent and subordinated to its own products, the sum of which is unreasonable and irrational. In the subsequent inversion, independent reason loses both its independence and its reasonableness, and manifests itself as something dependent and unreasonable, while products of this reason show up as the very seat of reason and autonomy. Reason no longer resides in the individual man and in his reason but outside the individual and outside individual reason. Unreason becomes the reason of modern capitalist society. The reason of society transcends the reason, powers and abilities of the individual, of the agent of Cartesian reason. Reason is transcendence. Cognition of this transcendence and of its laws is called science, subjecting to it is called freedom (freedom as 'the recognition of necessity'). Marx exposed these transcendental laws as a mystification of reason or as a mystified subject. This transcendence is the false subject whose force, power and reasonableness are nourished by the force, power and reasonableness of real subjects — socially acting people. Reason is the reason of an individual. The reasonableness of his reason is not, however, in its presuppositionlessness but rather in *including reasonable assumptions among the assumptions of his own reasonableness*. Therefore, while it lacks the immediate evidence of the Cartesian reason, reason is mediated by a *reasonably organized and reasonably shaped (social) reality*.

Dialectical reason not only seeks to know reality reasonably but also, and in particular, to shape it reasonably. But this had been the goal of rationalist reason as well. Where then do they differ? How did it happen that rationalist reason sought to shape reality reasonably yet did so unreasonably, so that the end product is a reality at once rationalized and irrational? Is the difference between dialectical and rationalist reason merely a methodological or an epistemological one, a result of substituting structural-genetic cognition, cognition of the concrete totality, for analytical-summative cognition? The starting point of rationalist reason is the atomized individual. Rationalist reason created modern civilization with its technology and its scientific achievements, but it also formed the rational individual, capable of exact scientific reasoning, as well as irrational forces, against which the 'rational individual' is powerless.

Rationalist reason thus officiates at the cradle both of modern science, as its foundation and its substantiation, and of the modern world with its rationalization and irrationality.[25] Rationalist reason forms a reality which it can neither grasp and explain nor organize in a consistent and rational fashion. This inversion is not a mystical transformation; it happens because the starting point of the *entire* process is the rationalist reason of an individual, i.e. both a particular historical form of reason, and the reason of a particular historical form of an individual. *This* reason must leave certain realities beyond the scope of reason: either because they cannot be *captured* by this reason and in this sense are irrational (the first meaning of irrationality), or because they cannot be *governed* and controlled by this reason, because they escape its rule and are irrational in this sense (the second meaning of irrationality).

This reason leaves aside something irrational (in the indicated two meanings of the word) and at the same time *forms* this irrational as a form of its own realization and existence. Rationalist reason assumed that the individual can 'use his reason for everything' and in this sense it opposed any authority and tradition. It wanted to investigate and know everything with its *own* reason. Apart from this positive aspect which is a *permanent feature* of modern thought, it also contained a negative aspect: a certain naiveté with which it ignored the fact that an individual is not only the subject who posits but is himself posited;* that as soon as it is realized, the reason of an atomised individual *necessarily* produces unreason because it takes itself as immediately given and does not include, in practice or in theory, the totality of the world. *Rationalization and irrationality are two incarnations of rationalist reason.* Rationalization of reality and the concurrent transformation of human reality into an objectual reality, as well as the irrationality and unreason of conditions which are at once impenetrable and ungovernable, all stem from the same foundation. Hence also the possibility of mistaking the rational (*racionální*) for the efficient (*racionelní*). If value judgements are excluded from science, and if science can rationally justify only the effectiveness of the means but not the appropriateness of the goals when dealing with human behavior (for otherwise it would lose its scientific character), then the influence of reason is limited merely to issues of action *techniques*. Furthermore, the overall issues of means, manipulation and techniques which pertain to the sphere of 'reason' become radically divorced from values and goals, i.e. from the subjective human world which is then abandoned to unreason, i.e. to irrationality. This conception appears both in Max Weber[26] and in the philosophical presuppositions of the mathematical and logical work of von Neumann and Morgenstern.[27] It considers as rational (in our terminology: efficient) such behavior which leads to an effective use of resources to goal achievement with minimum energy expenditure, or to maximum advantages. Science provides men with instructions on how to use resources efficiently and what means to employ in order to reach a given goal. However, it excludes discussions of the goal itself or of its justification and rationality. 'The rational character of our activity is gauged merely by the appropriateness of the means employed: goals are not subject to any purely rational evaluation'.[28]

Since the efficient and the irrational share a common origin, they can coexist in harmony, as manifest in the rationalization of the irrational and in the irrational consequences of rationalization. This concept of reason and this reality of reason equate reason with technology: they take technology as the perfect expression of reason and reason as the technique of behavior and action. Splitting scholarship into the sciences and the humanities, separating the methods of *erklären* and *verstehen*, as well as the recurrent naturalization and physicalization of social phenomena and the spiritualization of natural ones, all *manifest* with great clarity the cleaving of *reality*: the reign of rationalist reason is this cleft petrified. Human reality is divided

*A very unclear phrase in the Czech original.—Tr

both in theory and in practice between the sphere of the efficient, i.e. the world of rationalization, resources and technology, and the sphere of human values and meanings which in a paradoxical fashion become the domain of the irrational.

The unity of the capitalist world[29] is thus effected as a cleft between the world of calculation, manipulation, control, exact sciences, quantification, rule over nature, utility, in short: the world of objectivity, on the one hand, and the world of art, inner feelings, beauty, human freedom, religion, in short: the world of subjectivity, on the other hand. This is the objective ground which has time and again provoked attempts at an apparent or real reunification of the world or at complementing its one-sidedness: Pascal's 'logic of the heart' as a complement of Descartes' discursive method, *veritas aesthetica* as a complement of *veritas logica* (Baumgarten), transcendentalism which would overcome physicalism.[30]

Despite the heroic endeavor to explain everything rationally and to apply reason to everything,[31] classical rationalism of the 17th and 18th centuries produced a wave of real or apparent irrationalism. It grasped reason and rationality metaphysically and consequently failed to fulfil its own program. Very clear dialectical elements were, nevertheless, germinating even within the general metaphysical tendency, as demonstrated by the case of Leibniz. In turn, contemporary 'radical rationalism' of logical empiricists provokes an irrationalist reaction by simply *excluding* vast domains of reality from rational investigation and by abandoning them with voluntary defeatism to metaphysics and mythology. It is understandable why even non-Marxist philosophers who strive for a dialectical synthesis of scientific thinking and who critically continue in classical tradition, trying to formulate a modern dialectical rationalism, do not wish to share 'this pessimism which leaves to irrationalism and suggestion not only the humanities but all that concerns our actions, moral and political problems that transcend the purely technical sphere, i.e. that touch on philosophy'.[32] Rational scientism that excludes rational philosophy from science is necessarily complemented by irrational tendencies such as *Lebensphilosphie*, existentialism, neo-romanticism. Scientism and all manner of irrationalism are complementary products.

Metaphysical reason petrifies the rational and the irrational, grasps them as once-and-for-all given and immutable, and in this sense divides the *historically shifting* boundaries of man's *cognition* and the process of his *forming* reality into two ontological spheres: the existent of the rational and the existent of the irrational. On the contrary, the history of modern dialectics has demonstrated that dialectical reason abolishes these historical boundaries and that on behalf of man, on behalf of rationality in the broad sense, it gradually conquers 'areas' which metaphysical reason had considered an exclusive domain of irrationalism. Just as Hegel in his time found an ingenious answer to the historical alternative between rigid rational thought and irrational dynamism, an answer which amounted to a philosophical argument for dialectical reason, namely that 'there exists a higher type of rationality than that of abstract rigid thought',[33] so too are modern natural sciences and the materialist dialectical philosophy of the twentieth century consciously or spontaneously arriving at an adequate solution to the problem of the rational and the irrational in dialectical reason.

Dialectical reason is the universal and necessary process of cognition and of forming reality. It leaves nothing *outside* itself, and therefore becomes the reason both of science and thinking, and of human freedom and reality. The unreason of reason, and thus the historical limitation of reason, is in its denial of negativity. The reasonableness of reason is in that it assumes and anticipates negativity as its own product, in that it grasps itself as a continuing historical negativity, and thus knows that *its own activity is in postulating and solving contradictions. Dialectical reason does not exist outside reality nor does it leave reality outside itself. It exists only through realizing its reasonableness: that is, it forms itself as dialectical reason only insofar as it forms in the course of history a reasonable reality.*

The main features of dialectical reason can be summed up in the following points: (1) The historical character of reason as opposed to the transhistoricity of rationalist reason. (2) In contrast to the analytical–summative approach of rationalist reason which proceeds from the elementary to the complex and progresses from once-and-for-all determined starting points to the sum-total of human knowledge, dialectical reason proceeds from phenomena to the essence, from parts to the whole, etc., and conceives of progress in knowledge as a dialectical process of totalization which includes the reversibility of basic principles. (3) Dialectical reason is the capacity for rational thinking and cognition as well as a process of rationally forming reality, i.e. the realization of freedom. (4) Dialectical reason is negativity which places every completed step in cognition and in realization of man's freedom into a context of evolutionary totality where it transcends it, both in theory and in practice. It does not confuse the relative with the absolute but *grasps and realizes* the dialectics of the relative and the absolute in an historical process.

METAPHYSICS OF CULTURE

The Economic Factor

What is the economic factor and how did the belief in economic factors originate? In the course of metaphysical—analytical investigations, different aspects of the social whole are transformed into special autonomous categories. Individual moments of man's social activity — law, morals, politics, economics — are transformed in people's heads into autonomous forces which determine human activity. After isolating these individual aspects of the social whole and transforming them into abstractions, one studies their interconnections, such as the dependence of law on the 'economic factor'. This way of thinking turns products of man's social activity into autonomous forces which gain supremacy over man. Any synthesis of these metaphysical abstractions can therefore only be an external one, and any interconnection of the abstract factors a formal and mechanistically causal one. The factor theory was perhaps appropriate when prerequisites for a science of society were only just being developed; but the very successes of specialized social science research have resulted in substituting a superior scientific view — the synthetic investigation — for the factor theory.

We have followed almost word for word the argument of Labriola and Plekhanov who are credited with having studied the origin and the historical role of the factor theory. However profound was the distinction they made between the 'economic factor' and the economic structure (and we shall yet return to this distinction), their analysis is deficient in one point. According to both thinkers, the 'economic factor' and the belief in social factors were the result of *reflection*, a concomitant feature of underdeveloped *scientific thinking*.[34] Such conclusions deal only with the impact or with the consequences of factors but not with the problem of their *origin*. Decisive and primary is not the underdevelopment of scientific *thinking* or its limited, one-sided analytical form, but rather the disintegration of *social being*, the atomization of the capitalist society. 'Factors' are primordially products not of thinking and of scientific investigation but of a definite historical form of development: in the course of which artifacts of people's social activity become autonomous, in this form turn into factors, and traverse into uncritical consciousness as *autonomous forces* independent of man and his activity. We disagree with Plekhanov's and Labriola's *interpretation* of the origin of economic and other factors, and suggest that theirs is a one-sided approach smacking of Enlightenment. However, we completely accept their distinction between the economic factor and the economic structure. 'Does this mean that the economic factor and the economic structure are one and the same? Of course not, and it is quite curious that Mr Kareyev and his partisans have not understood this'.[35]

The distinction between the *economic structure* (a fundamental concept of Marxist materialism) and the *economic factor* (a much-used concept of vulgar sociological theories) offers the key to comprehending the central importance of political economy in the system of social sciences, and the priority of economics in the life of society.[36] The cardinal question, very important for grasping Marxism and its various concepts, is the following: Could pre-Marxist political economy have become the basis for a scientific, i.e. for a materialist conception of history? To comprehend the significance of economics both as *the economic structure of society* and as the science of the relations involved in it amounts to clarifying the very character of economics: economics is not a factor of social development, and the science of economics is consequently not a science of this factor. The critique which argues that the materialist theory of history holds only for the capitalist epoch because this is when materialist interests prevail and when the economy becomes autonomous (while Catholicism prevailed in the Middle Ages and politics in antiquity) demonstrates glaring lacunae in its grasp of Marx's theory. The prevalence of politics in antiquity, of Catholicism in the Middle Ages and of economics and material interests in modern times is explicable precisely on basis of the materialist theory, by elucidating *the economic structure* of each of these societies. Therefore when bourgeois ideology admits that material interests and the so-called economic aspect do play an important role in modern society, and benevolently concedes that Marxism has 'correctly' and 'inspiringly' pointed this out (even though being proverbially one-sided, it did not cover the whole truth), it lets its very own presuppositions fall victim to its own mystification. Its benevolence concerning Marxism is ridiculous. The prevalent role of the economic factor[37] observed by various bourgeois thinkers *before* Marx (Harrington, Madison, Thierry, etc.) itself calls for a materialist interpretation, i.e., it has to be interpreted on basis of the economic structure of capitalism and its peculiarities. The suspected autonomy of economics in the capitalist society, an autonomy that had not existed in previous societies, is an autonomy of *reified* social relations, and is therefore related only to one particular historical form of economics.

A different opinion has it that in an overall view of history, Marxism does not recognize the necessary prevalence of this or that sector of social life. The primacy of economics in the development of society is supposed to be only empirical and not necessarily inevitable, and it is supposed to disappear at that point in development at which the acquisition of material goods becomes a *secondary* matter, thanks to the great advance in production forces. In other words, economics plays according to this opinion a decisive role in relatively backward societies where due to underdeveloped production forces people have to devote *most* of their energies to problems of producing and distributing material goods. Economics is grasped exlusively in a *quantitative* sense, as one particular kind of human activity that is temporarily prevalent within the totality of this activity. Emancipating people from the quantitative domination of economic activity thus signifies the emancipation of society from the primacy of economics. But cutting down working hours, a prerequisite for emancipating people from the primacy of the economic factor, in no way eliminates the fact that people will be entering into certain social relations of production even in a free society, and that even then, production will have a social character. The *fetishism* of economics and the *reified* character of labor will disappear and exhaustive physical labor will be done away with. All this will allow people to devote themselves more to non-pro-ductive, i.e. non-economic activity. Nevertheless, *the economic structure will continue to maintain its primacy as the fundamental basis of social relations.* More precisely: *People will be emancipated from the supremacy of the economic factor only in one particular economic structure, i.e. a communist one.* We might point to the character of those classes which in past societies had been free from the immediate struggle for material goods and in this sense had not been under the supremacy of the economic factor. The *character* of these classes, the content and significance of their activity complete with the fact that it had been an unproductive activity were all consequences of the economic structure of their societies.

In his criticism of the factor theory, Kurt Konrad demonstrated that it is the fruit and the residue of a fetishist intuiting of society which mirrors social relations as relations among things. The factor theory turns social movement upside down. It considers isolated products of human objective or spiritual praxis to be 'agents' of social development, though in reality the only agent of social movement is man himself, in the process of producing and reproducing his social life.

Making the distinction between the economic structure, a category of Marxism, and the economic factor, a category of sociologism, is a prerequisite for scientifically substantiating and proving the primacy of economics in the life of society. The factor theory avers that one privileged factor – economics – determines all other factors: the state, law, art, politics, morals. In so doing it avoids the question of how the social whole, i.e. society as an economic formation, originates and is formed. It takes its formation for granted, as a given fact, as an uninvolved external form or arena in which the one privileged factor determines all the others. By contrast, materialist theory starts out from the opinion that *the social whole (the socio-economic formation) is formed and constituted by the economic structure. The economic structure forms the unity and continuity of all spheres of social life.*[38] Materialist monism – as opposed to all manner of pluralist theories – does not consider society to be a series or a cluster of factors, some of which appear as causes and others as effects. To face the choice between mechanical causality, where one factor is the cause and another the effect, and pluralist interaction, i.e. mere continuity which excludes *any* causality and substitutes functionality, assignation, etc. for it, is in itself a consequence of a *particular* view of reality. This view has first extracted certain isolated abstractions from social reality, promoted them to ontological existents (factors), and then backtracked and introduced these metaphysical constructs into various contexts, interactions or causal dependences. Naturally, the metaphysical point of departure necessarily leaves its mark on all this activity.[39] A metaphysical standpoint has been smuggled into the question itself.

Materialist monism considers *society* to be *a whole which is formed by the economic structure*, i.e. by the sum of social relations that people *in production* enter into with respect to *means of production*. It can provide a basis for a complete theory of classes, as well as an objective criterion for distinguishing between structural changes that affect the character of the entire social order, and derivative, secondary changes that only modify the social order without fundamentally altering its character. Contemporary apologies of capitalism (e.g. the opinion that class differences have been abolished in the most advanced imperialist countries) are based on theories which confuse the economic factor and the economic structure. We therefore feel it as no coincidence that the extensive apologetic literature concerning classes stems from Max Weber who considered the ability to dispose with property on the market to be decisive for class membership. This is an approach which completely wipes out the difference between the ownership of means of production, and the one hand, and of goods, on the other. In the place of the fundamental class dichotomy – between the exploiters and the exploited – Weber introduces an autonomous and therefore abstract scale of the propertied and the propertyless, of the wealthy and the poor, of those disposing and not disposing with property, etc. In other words: The concept of economics is reduced in this conception to the old 'factor theory' with economics taken as wealth, property, force of money, power of ownership, etc. This theory leads to the superficial polemical conclusion that an economically powerful individual need not be a real agent of power.[40] The construed one-sidedness of 'economic determinism' is countered with a pluralist determinism of economics, power, and social status. This is actually an opinion that *regresses* to the atomistic factor theory. Economics, power and social status constitute for Max Weber independent autonomous series that lead a transhistorical existence. In reality however (a) economic position, (b) social status with the hierarchy of social prestige, honor, respect, etc. and (c) the division of political power all enjoy a relative autonomy only *within and on the basis of* a particular socio-economic formation, in whose framework they function, interpenetrate and interact. The opinion that social status and political power are not 'in the last analysis' dependent on economics and on the economic structure of society, but rather constitute three independent, autonomous series, influencing one another, is an optical illusion, a result of grasping economics simplistically as the one factor around which other factors have to be arrayed in the interest of 'completeness'. It is true that ownership of money (*Geldbesitz*) is not in and of itself a status qualification, just as poverty is not in and of itself a disqualification. But even though property or poverty represent economic status rather then class membership, the concrete impact of this for social status and for politics will still *depend on the socio-economic structure*. For example, the problem of quixotism can be interpreted as one of transferring values such as status honor from the old, vanishing order in which they had functioned *normally* to a society whose structure and hierarchy of values are different. *Old* values function in it as extraordinary, and have an entirely *opposite impact or significance. The change in the functioning of certain values is not primordially a result of subjective evaluation but of an objective change in social relations.* Similarly with problems of power, of the power structure and of changes in it: they cannot be grasped on basis of the economic factor (of wealth, power of property, etc.) but only from the laws of this or that social formation's economic structure. To sum up, one might say: *The distribution of wealth ('economics'), the hierarchy and structure of power ('power'), and the gradation of social status ('prestige') are all determined by lawlike relations that in a given period of development stem from the*

economic structure of a social order. Questions arise as to how is power
distributed in a given society, who is the agent of power, how is it
executed — that is, questions concerning the nature of *the hierarchy of
power;* further, as to what is the scale and gradation of social prestige, who
receives the honors and how, who is the revered authority and the hero and
who the heretic and the 'devil', in other words, what is the character and
scale of *social status*; and finally, in what way is wealth distributed, how
does society break down into the propertied and the poor (or shall we say
the propertied and the less propertied) — that is, concerning *the distribution
of wealth.* Weber and his school consider all these problems to be
autonomous ones. Yet they all derive from the economic structure of social
formations, and only on this basis can they be rationally explained.

Emphasis on the *unity* of social reality formed by the economic structure
might of course become a hindrance to scientific investigation if this unity
were mistaken for a metaphysical identity, and if the concrete totality of
social reality were to degenerate into abstract wholeness. This explains how
contemporary sociology could have achieved certain positive results even
though it has abandoned the monistic methodological point of departure
and has switched to a detailed examination of particular areas or moments
of social reality for which it has created entire independent scientific
disciplines (sociology of power, sociology of art, sociology of culture,
sociology of knowledge, sociology of religion, etc.). In turn, *mere insistence*
on the correct — potentially correct, that is — point of departure will *in and
of itself, without realizing the truth of this starting point in its concrete
totality*, lead only to boorish repetitiveness, and will stagnate in a
metaphysical identity or in an empty totality.

Art and Its Social Equivalent

Philosophical questioning radically differs from walking around in circles.
But who is lost in circles and who is posing philosophical questions? Circular
reasoning operates with the naive unconscious idea that the confining circle
of questions is of its *own* making. The problems have been outlined, the
questions proposed, and reasoning now concentrates on refining its own
concepts. However, who was it who outlined and selected the problems?
Who drew the circle that constricts reasoning?

Arguments about realism and non-realism lead to recasting definitions
more precisely, to reforming concepts, to substituting words for other
words, but all this bustle is based on an unexpressed and unexamined
presupposition. People argue over the artist's attitude to reality, over the

means he has used to describe it, over the appropriateness, fidelity and artistic precision with which he has reflected this or that aspect of reality, taking it all the time tacitly for granted that the most obvious and most familiar thing, the thing least in need of any questioning and investigation, is none other than reality itself. Yet what is reality? How fruitful can arguments about realism and non-realism be if they clarify *secondary* matters while leaving the *cardinal* question in the dark? Does this discussion not require a 'Copernican turn' which would stand the whole up-side-down set of problems back on firm ground, clarify the cardinal question, and thus furnish the prerequisite for solving all others as well?

Every idea of realism or non-realism is based on a conscious or unconscious conception of reality. What is considered realism or non-realism in art always depends on what reality is and how it is conceived. *A materialist examination of the problem* therefore *begins* by positing this dependence as fundamental.

Poetry is not a reality of a lower order than is economics. It is an equally human reality, though of a different type and different form, with a different mission and significance. Economics does not beget poetry, directly or indirectly, mediately or immediately. Rather, man forms both economics and poetry as artifacts of human praxis. Materialist philosophy cannot buttress poetry with economics. Nor can it garb economics as the one and only reality into assorted less real or almost imaginary disguises such as politics, philosophy or art. Instead it has to ask the primary question about the origin of economics itself. He who takes economics as something given and further irreducible, as the ultimate original source of everything and the only real reality which cannot be questioned further, transforms economics into a result, a thing, an autonomous historical factor, and fetishises it in the process. Modern materialism is therefore a radical philosophy because it does not treat man's artifacts as the limit of analysis but penetrates to the roots of social reality, i.e. to man as the objective subject, to man as the being that *forms* social reality. Only on the basis of the materialist determination of man as the objective subject, i.e. as a being that uses natural materials to form a *new*, socio-human reality, in accordance with laws of nature and based on nature as an indispensible condition, can we *interpret* economics as the basic structure of man's objectification, as the master plan, the matrix of social relations, as the elementary level of human objectification and the economic base determining the superstructure. The primacy of economics is not the result of some of man's creations being more real than others but rather of *the central*

importance of praxis and work in the process of forming human reality.
Renaissance contemplations of man (and Renaissance *discovered* man and his
world for the modern era) began with work, conceived in the broad sense of
creating, i.e. as something that distinguishes man from beast and pertains
exclusively to man: God does not work, though he creates, but man both
creates and works. In Renaissance, creating and working were still united.
The new-born human world was as fresh and translucent as Botticelli's Venus
stepping out of a seashell in springtime. Creating is something exalted and
elevating. There is a direct connection between work as creating and the
elevating creations of work: creations point at their creator – man – who
stands *above* them, and testify not only to what he has become and has
achieved but to all that he can yet be. They annunciate his actual creativity
but even more so his infinite potentialities. 'All that surrounds us is our own
work, the work of men: all the houses, palaces, cities, marvellous buildings,
all over the country. They resemble the work of angels and yet are the work
of men . . . Seeing such marvels we understand that we can create even
better, more beautiful, more refined, more perfect things than hitherto
. . . .'[41]

Capitalism cuts this direct link, separates work from creating, creations
from their creators, and transforms work into uncreative, exhausting
drudgery. Creating is art whereas industrial labor is rote, something routine,
repetitive, and thus unworthy and self-devaluing. Man, in Renaissance the
creator and subject, sinks to the level of a creation and an object, to the
level of tables, machines, hammers. Having lost control over the material
world he had created, man loses reality itself as well. The real reality is now
the objective world of things and of reified human relationships. In
comparison with it, man appears as the source of mistakes, subjectivity,
imprecision and arbitrariness – in a word, as an imperfect reality. By the
19th century, the supreme reality no longer reigned in the heavens as the
transcendental God, the mystified idea of man and nature, but had
descended down to Earth as transcendental 'economics', the material
fetishised product of man. Economics turned into the economic factor.
What is reality and how is it formed? Reality is 'economics', and anything
else is a sublimation or a disguise of 'economics'. What then is economics?
'Economics' is the economic factor, i.e. that component of the *fetishised*
social being which has achieved autonomy and indeed supremacy over the
powerless disintegrated man, atomised in the capitalist society. In this
fetishised form or deformation it entered the consciousness of 19th century

ideologists, wreaking havoc as the economic factor, the primordial cause of social reality. The history of social theories records dozens of names, and we could add others, of people for whom economics had acquired this mysterious autonomy. These are the ideologists of the 'economic factor'. We wish to emphasize that Marxist philosophy has nothing in common with this ideology.

Marxism is no mechanical materialism that would reduce social consciousness, philosophy and art to 'economic conditions' and whose analytical activity would entail revealing the earthly kernel of spiritual artifacts. Materialist dialectics on the contrary demonstrates how a concrete historical subject uses his material—economic base to form corresponding ideas and an entire set of forms of consciousness. Consciousness is not reduced to conditions; rather, attention is focused on the process in which *a concrete subject produces and reproduces a social reality, while being historically produced and reproduced in it himself as well.*

The uncritical assignment of rigid and unanalysed intellectual phenomena to equally rigid and uncritically conceived 'social conditions', an approach so frequently attributed to Marxists and presented as all but the principle of their method, is in fact characteristic of a number of idealist authors. It serves them as a scientific interpretation of reality. Wildest idealism thus ends up hand in hand with the most vulgar materialism.[42] One of the most widespread instances of this symbiosis concerns the problem of romanticism. One section of the literature explains romantic poetry and philosophy on basis of the economic weakness of Germany, the impotence of the German bourgeoisie at the time of the French Revolution, or by the fragmentation of Germany and the backwardness of its conditions at the time. It seeks the truth about fixed rigid artifacts of the mind, which in this sense remain uncomprehended and external, in the conditions of a certain period. Marxism however – and this is its revolutionary contribution – was the first to propose that the truth of social consciousness is in social *being. Conditions, however, are not being.* Substituting 'conditions' for 'being' results in a number of other misconceptions: the idea that romanticism is merely a sum of props attributable to a particular historical *instance* of romanticism – such as the Middle Ages, an idealised people, phantasy, romanticised Nature, Desire, etc. – although in fact, romanticism continuously comes up with new props and discards old ones; the idea that romanticism differs from non-romanticism in that the one clings to the past and the other turns toward the future, although precisely romantic trends of the twentieth century have in fact proven that the future, too, is an important category of romanticism; the idea that romanticism and non-romanticism differ in that the one yearns for the Middle Ages while the other is attracted to antiquity, although in fact antiquity — and anything else, for that matter — can also be the subject of romantic longing.

This concept thus presents on the one hand conditions that form the content of consciousness, and on the other hand a passive consciousness, molded by these conditions. While the consciousness is passive and impotent, conditions are determining and omnipotent. What are these 'conditions', though? Omnipotence is not a necessary quality of 'conditions', just as passivity is not an eternal quality of consciousness. The antinomy between 'conditions' and consciousness is one of the different transitory historical forms of the subject-object dialectics which in turn is the basic factor of the dialectics of society.

Man does not exist without 'conditions' and is a social being only through 'conditions'. The contradiction between man and 'conditions', the antinomy between an impotent consciousness and omnipotent 'conditions', is an antagonism within 'conditions' themselves and a split within man himself. Social being is not equivalent to conditions, circumstances, or to the economic factor, all of which, taken in isolation, are deformations of that being. In certain phases of social development, man's being is cleft because the objective aspect of his being, without which he ceases to be man and turns into an idealistic vision, is separated from human subjectivity, activity, from his potentialities and possibilities. In this historical split, the objective side of man is transformed into alienated objectivity, a dead, inhuman objectivity (into 'conditions' or the economic factor), and human subjectivity is transformed into a subjective existence, squalor, need, emptiness, mere abstract possibility, a yen.

Man's social character is evident not only through his being nothing without an object, but particularly in that he manifests his reality in *objective activity*. In producing and reproducing social life, i.e. in forming himself as a socio-historical being, man produces the following:

(1) material goods, the material-sensory world based on work,

(2) social relations and institutions, the sum of social conditions, and,

(3) based on these, he produces ideas, emotions, human qualities and corresponding human senses.

Without a subject, these social products of man would be senseless, while without material means and objective creations, the subject would be a mere specter. *The essence of man is the unity of objectivity and subjectivity.*

Man has formed himself on basis of work, in work and through work, not only as a thinking being, qualitatively different from all other higher animals, but also as the only being in the universe we know of capable of forming reality. Man is a component of nature and is himself nature. At the same time, though, he is a being which, having mastered both 'external' and his own natures, forms a *new* reality in nature, one that is irreducible to the latter. The world that man constructs as a socio-human reality stems from conditions independent of man, and is unthinkable without them. Yet it represents a different quality, irreducible to these conditions. Man stems from nature and is a part of it even as he transcends it. He relates freely to his creations, steps away from them, questions their meaning and questions his own place in the universe. He is not closed within himself and his world. Because he forms a human world, an objective social reality, and thus is able to transcend its situation, conditions and assumptions, man grasps and interprets the extra-human world as well, the universe and nature. Man can penetrate the mysteries of nature only because he *forms* a human reality. Modern technology, experimental laboratories, cyclotrons and missiles all disprove the suggestion that cognition of nature is based on mere contemplation.

Human praxis thus appears in yet another light: as the arena for the metamorphosis of the objective into the subjective and of the subjective into the objective. It is the 'active center' in which human intentions are realized and laws of nature discovered. Human praxis unites causality and purposiveness. And if we take human praxis as a fundamental social reality, it becomes obvious that based on praxis, human consciousness fulfills two indivisible basic functions: of registering and of projecting, of fact-finding and of planning. It is at once a reflection and a project.

The dialectical character of praxis imprints an indelible stamp on *all* human creations, including art. A medieval cathedral is an expression or an image of the feudal world but at the same time is a constructive element of it. The cathedral not only artistically reproduces medieval reality, it also produces it artistically. *Every work of art has an indivisible two-fold character: it expresses reality but also forms it. It forms a reality that exists neither beyond nor before the work itself but strictly in the work only.*

The patricians of Amsterdam are reported to have angrily rejected Rembrandt's 'Night Watch' (1642) in which they did not recognize themselves and which impressed them as distorting reality. Is reality truthfully known only when one recognizes oneself in it? This suggestion would assume that man knows himself, knows what he looks like and who he is, that he knows reality and can tell what reality is, *independently* of art and philosophy. But from where does man know all this, and from where comes the certainty that what he knows is indeed reality itself and not merely *his* idea of it? The patricians defended their idea of reality against the reality of Rembrandt's work and thus equated their prejudices with reality. They believed reality was contained in their ideas and thus that their ideas were reality. It followed logically that an artistic expression of reality should translate their ideas into the language of sensory artistic painting. Reality was known and the artist should only depict and illustrate it. But a work of art does not depict *ideas* of reality. As work and as art, it both depicts reality and *forms* it, simultaneously and inseparably: the reality of beauty and art.

Traditional interpretations of the history of poetry, philosophy, painting, music, etc., recognize that all *great* artistic and intellectual currents emerged in a struggle with ingrained ideas. But why is this so? We hear references to the weight of prejudice and tradition. 'Laws' have been invented according to which artifacts of the mind evolve in an historical alternation of two 'eternal' types (classicism and romanticism) or swing as a pendulum from one extreme to another. These 'interpretations' interpret nothing. They only obfuscate the problem.

Assumptions of contemporary science are based on the Galilean revolution. Nature is an open book and man can read in it, providing of course that he has *mastered the language* in which it is written. Since the language of nature is the 'lingua mathematica', man cannot *interpret* nature scientifically and *control* it practically unless he has mastered the language of geometrical figures and mathematical symbols. A scientific understanding of nature is denied to him who has not mastered mathematics. For him, nature (that is, nature in *one* of its aspects) is mute.

What is the language in which the book of the human world and of socio-human reality is written? How does this reality disclose itself, and to whom? If the socio-human reality knew its own reality, and if naive everyday consciousness knew it, philosophy and art would turn into an inessential luxury, recognized or rejected according to momentary need. Philosophy and art would merely be repeating in a conceptual language of ideas or in a metaphorical one of emotions something that would have been known even without them and that *for man* would exist independently of them.

Man seeks to grasp reality but frequently merely 'gets hold of' its surface or false image. How then does reality disclose itself in its own reality? How does the truth of human reality reveal itself to man? Man learns about partial areas of the socio-human reality and establishes truths about them through specialized sciences. He has furthermore two different 'means' that lead him to cognition of human reality *as a whole* and to disclosing the truth of reality *in its own reality*: philosophy and art. *This* is what the special position and the special mission of art and philosophy are based on. Because of their vital and indispensable function, art and philosophy are insubstitutable and irreplaceable. They are, Rousseau would say, inalienable.

Reality discloses itself to man in great art. Art in the proper sense of the word is at once demystifying and revolutionary because it ushers man away from his ideas and prejudices about reality, and into reality itself and its truth. True art and true philosophy[43] reveal the truth of history: they confront mankind with its own reality.[44]

Which is the reality that manifests itself to man in art? Is it a reality which he *already* knows and now wants to appropriate in a *different* manner, i.e. through sensory intuition? Suppose Shakespeare's plays were indeed 'nothing but'[45] an artistic rendering of class struggle in the era of primitive accumulation, and that a Renaissance palace was indeed 'nothing but' an expression of the emerging class power of the capitalist bourgeoisie. The question that would then arise is, why do these social phenomena which exist in and of themselves independently of art have to manifest themselves in art *once again*? And in a form which no less disguises their real character, and thus in a way both conceals and reveals their true essence? This conception assumes that the truth expressed in art is attainable also by a different path; the sole difference being that while art presents this truth 'artfully', in graphic sensory images, in some other form this truth would be less *impressive.*

The Greek temple, the medieval cathedral and the Renaissance palace all express reality but they also simultaneously *form* it. Of course, they form more than the antique, medieval or Renaissance reality, more than the architectonic elements of these societies. As *perfect works of art,* the reality they form is one that transcends the historicity of their respective worlds. This transcendence lays bare the *specificity* of their reality. The reality of a Greek temple is different from that of an antique coin. The latter lost its reality with the demise of the antique world. It is invalid and does not function as a means for payment or for hoarding. With the demise of the historical world, functional elements lose their reality too: the antique temple loses its immediate social function as a place for worship and for religious ceremonies, the Renaissance palace is no longer the visible symbol of power and the actual seat of the Renaissance prince. But even with the demise of the historical world and with the passing of their social functions, the antique temple and the Renaissance palace still do not lose their artistic value. Why is this? Do they express a world which survives despite having disappeared in its own historicity? How does it survive and in what? As a sum of conditions? As material worked over by people who impressed their own character into it? A Renaissance palace points to an entire Renaissance world: extrapolating from a palace, one can decipher the contemporary man's attitude to nature, the degree of realization of human freedom, the organization of space, the expression of time and the conception of nature. However, a work of art expresses an entire world only insofar as it *forms* it. It forms a world insofar as it discloses the truth of reality, insofar as reality speaks out through the work of art. In a work of art, reality addresses man.

We started with the opinion that to examine both art's relationship to reality and the derived concepts of realism and non-realism, one has to answer the question: What is reality? On the other hand, the very analysis of works of art has also led to the main question, the main subject of our consideration: What is socio-human reality and how is it formed?

If the relation of social reality to a work of art were considered exclusively in terms of the conditions and historical circumstances which determined the work's genesis, then the work itself and its artistic character would acquire an extra-social character. If that which is social were predominantly or exclusively fixed in the form of reified objectivity, then subjectivity would be grasped as something extra-social, as a fact which is not formed and constituted by social reality, though it is conditioned by it. If the relation of social reality to a work of art were conceived as conditions of the times, as the historicity of circumstances, or as the social equivalent, the monism of materialist philosophy would collapse. Its place would be taken by a dualism of conditions and people: conditions would outline tasks, people would react to them. In modern capitalist society, the subjective moment of social reality has been severed from the objective one, and the two aspects confront each other as independent substances: as pure objectivity on the one side and reified objectivity on the other. Hence a double mystification: the automatism of conditions, and the psychologization and passivity of the subject. But social reality is infinitely more variegated and concrete than conditions and circumstances, precisely *because it includes human objective praxis* which forms these conditions and circumstances. Circumstances are the fixed aspect of social reality. The moment they are severed from human praxis, from man's objective activity, they become something rigid and uninspired.[46] 'Theory' and 'method' attempt to causally conjoin this uninspired rigidity with 'the spirit', with philosophy and poetry. The result is vulgarization. Sociologism reduces social reality to conditions, circumstances and historical determinants which, when deformed in this way, assume the form of natural things. The relations between 'conditions' and 'historical circumstances' conceived in this way, and philosophy and art, can *in principle* be none other than mechanistic and external. Enlightened sociologism endeavors to eliminate this mechanicism by introducing a complicated hierarchy of real or artificial 'mediating links' (and 'economics' is then connected with art only 'indirectly'), but this amounts to the toil of Sisyphus. For materialist philosophy which has introduced the revolutionary question: *How is social reality formed?*, this reality exists not only in the form of 'objects', conditions and circumstances, but above all as the objective activity of man who himself forms conditions as objectified components of human reality.

According to sociologism which, characteristically, substitutes conditions for social being, the human subject reacts to conditions as they change. He is an immutable set of emotional and intellectual abilities which capture, study and depict these conditions in arts and sciences. As conditions change and unfold, the human subject goes along and takes snapshots of them. He becomes a recorder of conditions. Sociologism tacitly assumes that while economic formations alternate throughout history, while thrones fall and revolutions prevail, man's ability to 'perceive' the world has remained unchanged since antiquity.

Man perceives and appropriates reality 'with all his senses', as Marx has stressed, but the very senses which reproduce man's reality are themselves a socio-historical product, too.[47] Man has to develop a particular sense if objects, events and values are to have sense for him. For the man who lacks such a sense, people, things and creations also lack sense and are senseless. Man *exposes* the sense of things by *forming* a human sense for things. A person with developed senses has consequently a sense for everything human, while a person with underdeveloped senses is closed to the world and 'perceives' it not universally and totally, sensitively and intensively, but one-sidedly and superficially, from the viewpoint of his own 'world', which is a one-sided fetishised segment of reality.

We do not criticise sociologism for having concentrated on conditions and circumstances, *in order to* interpret culture, but for not having grasped the significance of conditions by themselves, or in their *relation* to culture. Conditions *outside* history, conditions *without* a subject are not .only a petrified and mystified artifact but also lack all *objective sense*. 'Conditions'

in this form lack what is most important even from the methodological perspective — namely a proper objective meaning. Instead, they *acquire* a false sense that depends on the opinions, reflection and education of the scholar.[48] Social reality ceases to be for research what it is objectively, i.e. a concrete totality, and disintegrates into two independent heterogenous wholes which 'method' and 'theory' then strive to unite. The break-up of the concrete totality of social reality leads to petrifying conditions on one pole, and the spirit, psyché and the subject on the other pole. Conditions are then either passive and are set in motion and given sense by the spirit, psyché, or by the active subject in the form of an *'élan vital'*, or else they are active and become the subject themselves. Psyché or consciousness then has no other function than to examine, in an exact or in a mystified way, the scientific laws of these conditions.

It has been frequently observed that Plekhanov's method fails in the study of problems of art.[49] Its failure is manifest both in its non-critical acceptance of ready-made ideological constructs for which it then seeks an economic or a social equivalent and in the conservative rigidity with which it fences off its own path to comprehending modern art, considering impressionism to be the last word of 'modernism'. It seems, however, that the theoretical and philosophical origins of this failure have not been sufficiently examined. Plekhanov never overcame the dualism of conditions *versus* psyché because he never fully comprehended Marx's concept of praxis. Plekhanov quotes Marx's *Theses on Feuerbach* and notes that to a certain degree they contain the program of modern materialism. If Marxism does not want to concede that in some spheres idealism is stronger, it has to be able according to Plekhanov to present a materialist explication of *all* aspects of human life.[50] After this introduction, Plekhanov presents his own interpretation of Marx's concepts of 'human sensory activity', praxis, and subjectivity: 'The subjective aspect of human life is precisely the psychological one: "human spirit", emotions and ideas of people'.[51] Plekhanov thus distinguishes between psychology, psychic states, or the state of the spirit and morals, emotions and ideas, on the one hand, and economic conditions on the other hand. Emotions, ideas, the state of the spirit and morals are 'materialistically clarified' when clarified on the basis of economic history. It is perfectly clear that Plekhanov parts ways with Marx in the *cardinal point* at which Marxist materialism has succeeded in transcending both the *weaknesses* of all previous kinds of materialism, and the strong points of *idealism*: that is, in its grasp of the subject. *Objective praxis, Marx's most important discovery, consequently entirely drops out of the materialist conception of history.* Plekhanovist analyses of art fail because they are based on a concept of reality which lacks the constitutive elements of objective human praxis. It lacks the 'human sensory activity' which cannot be reduced to 'psyché' or to the 'spirit of the times'.

Historism and historicism

Marx's famous fragment on antique art shares the fate of many a brilliant thought: the sediment of commentaries and the self-evidence of daily references to it have obscured its true sense.[52] Was Marx investigating the meaning and the timeless character of antique art? Was he intending to solve problems of art and beauty? Is the passage in question an isolated expression or is it bound with other opinions of the author? What is its proper sense? Why do those commentators fail who consider exclusively its literal immediacy and take it for an invitation to resolve the question of Greek art's ideal character? And why do also those interpreters fail who consider Marx's immediate answer as satisfactory, without pausing to wonder why the manuscript abruptly ends in the middle of developing an idea which is left incomplete?

In this fragment which otherwise deals with the method of political economy, the methodology of social sciences and with problems of the materialist concept of history, considerations of art are of secondary importance. Marx is not specifically investigating the Greek epos, but uses it as an example for solving other, *more general* problems. He focuses attention not on explaining the ideal character of antique art but on formulating problems of *genesis* and *validity*: the socio-historical *constraints* of art and of ideas are not identical with their *validity*. The main issue is not problems of art but the formulation of one of the cardinal questions of materialist dialectic: the relationship between genesis and validity, conditions and reality, history and human reality, the temporal and the eternal, between relative and absolute truth. To solve a problem, one must first formulate it. Outlining a problem is of course something other than limiting it. To outline, to formulate a problem means to trace and determine its internal relations with other problems. The main problem concerns not the ideal character of antique art but is more general: How and why does a work of art outlive the conditions in which it had originated? In what and why do Heraclitus' thoughts survive the society in which they were developed? Where and why does Hegel's philosophy outlive the class whose ideology it had then formed? The actual question is a general one.

Only in the light of this general formulation can a specific question be grasped and solved. And on the contrary, the general problem of absolute and relative truth, of genesis and validity, can be exemplified on basis of the fully comprehended specific problem of antique art.[53] The problems of a work of art are to lead us to the problems of the eternal and the temporal, of the absolute and the relative, of history and reality. A work of art — and in a sense works in general, including those of philosophy and science — is a complex structure, a structured whole which conjoins diverse elements in a dialectical unity: ideas, themes, composition, language.[54] The relationship of a work to social reality cannot be adequately dealt with by declaring that a work is a structure of meanings which is open toward social reality and is *determined* by it both as a whole and in its individual constructive elements. Conceiving of the relation of the work to social reality as one of the determinants to what is determined would *reduce* social reality merely to social conditions, i.e. to 'something' that is related to the work only as an *external* prerequisite and as an external determinant.[55] The work of art is an integral component of social reality, a constructive element of this reality and a manifestation of man's social-intellectual production. In order to comprehend the character of a work of art it will not suffice to have a 'sociology of culture' deal with its social character and with its relationship to society, to examine its socio-historical genesis, impact and reception, or to have historical research investigate its biographical and socio-biographical aspects.

A work of art is indubitably socially determined. Uncritical thinking, however, reduces this relationship to the only connections between social reality and art, and thus distorts the character of both. The thesis about social determinism tacitly assumes that social reality remains *outside* the work. The work thus effectively turns into something extra-social, it does not constitute social reality, and thus has no *internal* relation to social reality. An analysis of the work could deal with the social determinism of the work separately, in a general introduction or an appendix, it could be placed before the brackets, as it were, but it would not enter the *actual* structure of the scientific *analysis*; indeed, it would not even belong there. In this relationship of mutual externality, both social reality *and* the work itself degenerate: if the work, a certain structure of meanings, does not enter the analysis and investigation of social reality, then social reality turns into a mere abstract framework or into general social determinism: concrete totality turns into a false totality. If one does not investigate the work as a structure of meanings whose concreteness is grounded in its existence as a moment of social reality, and if one admits determinism as the only 'link' between the work and social reality, then the work which is a relatively autonomous structure of meanings changes into a structure that is absolutely autonomous: concrete totality turns, again, into false totality. Two different meanings are hidden in the thesis of social determinism of the work. First, social determinism means that social reality is related to the work as a deistic God-Mover would be; it gives the first impulse, but once the work is created, it changes into a spectator who observes the autonomous development of his creation without influencing its fate any further. Second, social determinism means that the work is something secondary, derivative, mirrored, whose truth is not contained in the work itself but is outside it. If the truth of the work of art is not in the work itself but in conditions, it will be necessary to know all about these conditions in order to comprehend the work. Conditions are supposed to be the reality the work reflects. But in and of themselves, conditions are not reality; they are reality only insofar as they are the realization, fixing, and development of the objective praxis of man and his history, and insofar as they are grasped as such. The truth of a work (and for us, a work is always a 'real' work of art or of philosophy, as opposed to 'writings') is not in the situation of the times, in social determinism or in the historicity of conditions, but in socio-historical reality as the unity of genesis and replicability, and in the development and realization of the subject—object relationship as a specifically human existence. *The historism of social reality is not the historicity of conditions.*

Only now have we arrived at the point from which we can return to the original question: How and why does a work outlast the conditions in which it had originated? If the truth of a work is in the conditions, it survives only insofar as it is a *testimony* to these conditions. A work testifies to its times in two senses. First, by simply looking at a work we recognize which era it belongs to, which society engraved its mark on it. Second, we look at the work seeking the testimony it offers about its time and conditions. We take it as a document. In order to examine the work as a testimony to its times or as a mirror of its conditions, we first have to know these conditions. Only after comparing the conditions with the work itself can we decide whether the work mirrors its era in a straight or a crooked way, whether it testifies to its times truly or not. But every cultural creation fulfils the function of a testimony or document. A cultural creation that mankind looks upon *exclusively* as a testimony is not a work. It is a specific quality of works that they are not primarily or exclusively a testimony to their times. They do of course testify to the time and conditions of their genesis *as well*; but *apart from this* they are (or are in the process of becoming) constitutive elements of the existence of mankind, of classes and nations. Characteristic of works is not *historicity*, that is, 'bad uniqueness' and irreplicability, but *historism*, i.e. the capacity for concretization and survival.

By outlasting the conditions and the situation of its genesis, a work proves its vitality. It lives as long as its influence lasts. The influence of a work includes an event that affects both the consumer of the work and the work itself. What happens to the work is an expression of what the work is. That something happens to the work does not mean that it is abandoned to the play of the elements. On the contrary, it means the internal power of the work is realized over time. In the course of this concretization the work acquires different meanings. We cannot always say in good conscience that *every one* of them had been intended by the author. While creating, the author cannot foresee all variants of meanings and all the interpretations that will be imputed to his work. In this sense the work is independent of the author's intentions. On the other hand, though, this independence and autonomy are fictitious: the work is a work and lives as a work because it *calls for* interpretations and because it has an *influence* of many meanings. What are the grounds for the *possibility* of concretizing the work, for it acquiring various historical forms during its 'lifetime'? Clearly there has to be *something* in the work that makes this effect possible. There exists a certain span within which concretizations of the work are conceived as concretizations of *this* particular work. Beyond the limits of this span, one talks about distortions, lack of comprehension and about subjective interpretations of the work. Where is the borderline between an authentic and an inauthentic concretization of a work? Is it contained in the work itself or is it outside the work? How does a work that lives *only* in and through individual concretizations outlive every particular one of them? How does it manage to slough them off one after another, demonstrating its independence on them? The life of a work points beyond the work itself, at something that transcends it.

The work's life is incomprehensible from the work itself. If the work's influence were its property as radiation is a property of radium, then the work would live, i.e. exercise an influence, even when 'unperceived' by a human subject. The influence of a work of art is not a physical property of objects, books, paintings or sculptures, i.e. of natural or artificial objects. It is a specific mode of *existence* of the work as of a *socio-human reality*. The work lives not in the inertness of its institutional character or thanks to tradition, as sociologism would have it[56] but through totalization, i.e. reviving. The work's life is not the result of its autonomous existence but of the *mutual interaction of the work and mankind.* The work has a life because

(1) the work itself is infused with reality and truth, and because of

(2) the 'life' of mankind, i.e. of a producing and perceiving subject.

Every component of socio-human reality has to demonstrate this subjec-tive–objective structure in one form or another.

The life of a work of art can be conceived as a manner of existence of a *partial* structure of meanings, integrated in some way in the total structure of meanings – in socio-human reality.

A work that has outlasted the time and conditions of its genesis is frequently credited with the quality of timelessness. Is temporality perhaps something that gives in to time and becomes its prey? And conversely, does timelessness overpower and subjugate time? Timelessness of a work would literally mean its existence without time. The idea of a work's timelessness cannot, however, rationally cope with two basic problems: (1) How can a work, 'timeless' in character, originate in time? (2) How can one proceed from the timeless character of a work to its temporal existence, i.e. to its concretization? Conversely, the key question for every anti-Platonic concept is this: How can a work generated in time *acquire* a 'timeless' character?

What does it mean to say that a work withstands time or that it survives bad times? Is it resistant to decay and destruction? Or perhaps does the work cease to exist altogether as it resists time, and places time *outside* itself, as something external? Is eternity the exclusion of time and is timelessness the arresting of time? The question, 'What does time do to a work?' can be answered by another question: What does a work do with time? We arrive at the conclusion, paradoxical at first, that the timelessness of a work is in its temporality. To exist means to be in time. Being in time is not movement in an external continuum, but temporality, i.e. the realizing of a work in time. The timelessness of a work is in its temporality as activity. The timelessness of a work does not mean its permanence outside time or without time. Timeless permanence would amount to a stupor, to loss of 'life', of the ability of the work to set itself in time. The greatness of a work cannot be gauged by its reception when it first appears. Great works have been rejected by their contemporaries, others have been recognized as seminal immediately, yet others 'laid on the shelf' for dozens of years before 'their time' came. Whatever happens to a work is a form of what the work is. The rhythm of its 'temporality' depends on its nature: whether its message holds for every time and every generation, whether it has something to offer only at certain times, or whether it must first 'hibernate' in order to be revived later. This rhythm of reviving and of temporality is a constitutive element of the work.

It is a curious coincidence that adherents of historical relativism converge with their opponents, with advocates of natural rights, at a central point: both schools eliminate history. The basic thesis of historicism, that man cannot transcend history, as well as the polemical assertion of rationalism, that man has to transcend history and arrive at something metaphysical, something that would guarantee the truth of knowledge and morality, both share the assumption that history is variability, unique irreplicability and individuality. For historicism, history breaks up into the transience and temporality of conditions concatenated by no historical continuity *of their own* but only by a *transhistorical* typology, the explicative principle of human spirit, by a regulating idea that introduces order into a chaos of particulars. The formula that man cannot step outside of history indicates the impossibility of achieving objective truth. This, however, is an ambiguous formula, since history is more than historicity, temporality, transience and irreplicability which exclude the absolute and the trans-historical, as historicism would have it. Equally biased is the opinion that history as a happening is something insubstantial because in all of its metamorphoses, and thus *behind* history, there endures something trans-historical, absolute, something that the course of history cannot affect. History is external variability performed on an unvarying substance. The absolute that existed before and exists above history is also pre-human, for it exists independently of man's praxis and being. If the absolute, the universal and the external, is unvarying and if its permanence is independent of variability, then history is a history only in appearances.

Unlike the relativism of historicism and unlike the ahistorism of the concept of natural rights, dialectics considers nothing to be absolute and universal: be it prior to history and independently of it, or at the end of history as its absolute and final design. Rather, the absolute and the universal are *formed in the course of history*. Ahistorical thinking knows the absolute only as non-historical, and thus as eternal, in the metaphysical sense. Historicism culls the absolute and the universal out of history altogether. In distinction from both, dialectics considers history to be a unity of the absolute in the relative and of the relative in the absolute, *a process in which the human, the universal, and the absolute appear both in the form of a general prerequisite and as a specific historical result.*

History is a history only because it includes both the historicity of conditions and the historism of reality, because it contains ephemeral historicity which recedes into the past and *does not return*, as well as historism, i.e. the formative of that which endures, the self-formative and the creative. Man is always an historical being which never exits from the sphere of history. He thus stands, as a real possibility, above every act or circumstance in history and can set standards for evaluating it.

What is universally human, 'ahistorical' or common to all phases in history, does not exist *independently*, in the form of an immutable, eternal, transhistorical substance. It exists at once as the universal *condition* of every historical phase and as its specific *product. The universally human is reproduced in every epoch as a particular outcome, as something specific.* [5 7]
Historicism, in terms of historical relativism, is itself a product of a reality split between a transient, emptied, devalued facticity, and a transcendental existence of values *outside* reality. At the same time, though, historicism ideologically fixes this split. Reality breaks down into the relativised world of historical facticity and the absolute world of transhistorical values.

Yet what is that transhistorical value that either never becomes a part of conditions or else outlasts them? The belief in transcendental values of a transhistorical character suggests that the real world has been emptied and de-valued, that concrete values have disappeared from it. This world has become valueless, while values have occupied an abstract world of transcendence and moral obligation.

The absolute, however, is not divorced from the relative. It is rather 'composed' of the relative or, more precisely, is *formed* in the relative. If everything is subject to change and extinction, and if all that exists exists *only* in a certain time and a certain space, transience being its only quality, then the speculative theological question concerning the *sense* of the temporal and transient must remain an *eternal* and eternally unanswered question. The question concerning the relation of the relative and the absolute in history is dialectically formulated thus: How do historical degrees of mankind's evolution *turn into* transhistorical elements of the *structure* of mankind, i.e., of human nature? [5 8] How do genesis and evolution interconnect with structure and human nature? Classes, individuals, epochs and mankind itself have struggled to become conscious of *their own* practical–historical problems, in different formations of human consciousness. As soon as these are formed and formulated, they turn into components of human consciousness, that is into finished forms through which *every* individual can experience, be conscious of and realize problems of all mankind. Unhappy consciousness, tragic consciousness, romantic consciousness, Platonism, Macchiavellism, Hamlet, Faust, Don Quixote, Josef Schweik and Gregor Samsa are all historically *generated* forms of consciousness or ways of human existence. Their classical form was created in some *particular*, unique and irreplicable epoch but, once created, predecessors turn up in scattered fragments from the past, if only as comparatively crude attempts. As soon as they are created and are 'here', these classical expressions occupy a distinct place in history because they themselves form history and acquire a validity independent of the original historical conditions of their genesis. *Social reality as human nature is inseparable from its products and from forms of its existence. It does not exist other than in the historical totality of these products which, far from being external and accessory 'things', reveal and indeed retroactively form the character of human reality (of human nature).* Human reality is not a pre-historical or a transhistorical and unvarying substance. It is formed in the course of history. *Reality is more than conditions and historical facticity; but neither does it ignore empirical reality.* The dualism of transient and emptied empirical facticity on the one hand, and the spiritual realm of ideal values rising independently above it on the other hand, is the mode in which a particular *historical* reality exists: the historical reality *exists in this duality*, and its entirety consists of this split. Idealistically hypostatizing this historical form of reality leads to the conclusion that the world is cleaved into a true intransient *reality* of values and a false 'reality' or facticity of transient conditions.[59]

The only reality of the human world is the unity of empirical conditions, complete with the process of forming them, on the one hand, and of transient or living values and their formative process, on the other. The particular *historical* character of reality determines whether this unity is realized as a harmony of incarnated values, that is through conditions *infused* with values, or as a split between empty, invalidated empiricism and ideal transcendental values.

Reality is 'higher' than are the conditions and historical forms of its own existence. That is, reality is not the chaos of events or of fixed conditions but rather the unity of events and their *subjects*, a unity of events and the *process of forming them*, a practical–spiritual ability to transcend conditions. The ability to transcend conditions allows for the *possibility* to proceed from opinion to cognition, from *doxa* to *epistémé*, from myths to truth, from the accidental to the necessary, from the relative to the absolute. It is not a step out of history but an expression of the specificity of man as an event-formative and history-formative being; man is not walled into the animality and barbarism of his race, prejudices and circumstances,[60] but in his onto-formative [*seinsbildenden* German—Tr.] character (as praxis) he has the ability to transcend toward truth and universality.

As one of the ways of overcoming the temporary and the momentary,[61] human history is more than the ability to *store* and recall, i.e. to draw ideas, impressions and feelings out of the storage room of semi-oblivion or of the subconscious. It is also a particular active structure and organization of human consciousness, of knowledge. It is an *historical* ability and an *historical* structure because it is based not only on an historically evolving sensory—rational 'equipment' of man. It can draw past things out *into the present* and transcend the temporary because man does not leave the past behind as some discarded object. Rather, the past enters into his present and constitutes his present in form of a formative and self-formative human nature. Historical periods of human development are not empty casts from which life would have evaporated *because of* mankind having reached higher forms of development. Rather, they are continually incorporated into the present through praxis — the creative activity of mankind. This process of incorporation is at the same time a critique and an appreciation of the past. The past which is concentrated in the present (that is, abolished in the dialectical sense) shapes human nature: which is a 'substance' that includes objectivity as well as subjectivity, material relations and reified forces as well as the ability to 'see' the world and to explicate it in different subjective modes (i.e. scientifically, artistically, philosophically, poetically, etc.)

The society that gave birth to the genius of Heraclitus, the era in which Shakespeare's art was generated the class in whose 'spirit' Hegel's philosophy was developed, have all *irretrievably* vanished in history. Nevertheless, the 'world of Heraclitus', the 'world of Shakespeare', and the 'world of Hegel' continue to live and exist as living moments of the present[62] because they have enriched the human subject *permanently*.

Human history is an incessant totalization of the past, in the course of which human praxis incorporates and thus revives moments of the past. In this sense, human reality is *not only the production of new but also a* — critical and dialectical — *reproduction of the old*. Totalization is the process of production and reproduction, of reviving and rejuvenating.[63]

The capacity for and the process of totalization are at all times both a prerequisite and an historical result: the differentiated and universalized capacity of man's perception to admit equally as artistic treasures works of antiquity, creations of the Middle Ages, and the art of ancient nations is an *historical* product that did not exist and would have been unimaginable in any medieval or slave society. Medieval culture could not have revived (totalized and integrated) antique culture or the culture of 'pagan' nations without exposing itself to the danger of disintegration. On the contrary, progressive modern culture of the twentieth century is a universal culture in its own right, with a high capacity for totalization. While the medieval world was blind and closed to expressions of beauty and truth of *other cultures*, the modern view of the world is by contrast based on universality, on the ability to absorb, perceive, and appreciate expressions of most diverse cultures.

NOTES

[1] See Translator's Note, p. 92.

[1a] 'The Latin word *cura* is ambiguous . . . Man's lifelong *cura* includes an earthly, pedestrian element oriented toward the material, but also an element aspiring toward God.' K. Burdach, 'Faust und die Sorge', *Deutsche Vierteljahrsschrift für Literaturwissenschaft und Geistesgeschichte*, 1923, p. 49.

[2] The critique that sees in *Being and Time* the patriarchal world of backward Germany has fallen for the mystification of Heidegger's *examples*. Heidegger, however, is describing problems of the modern twentieth century capitalist world which he exemplifies – quite in the spirit of romantic disguising and concealing – by the blacksmith and forging. This chapter is not an analysis of Heidegger's philosophy but of 'care' which represents the reified moment of praxis, as does the 'economic factor' and the 'homo oeconomicus'.

[3] S. L. Rubinstein, *Printsipi i Put'i Razvit'ia Psikhologii*, Moscow, 1959, p. 204. In this section, the author polemicises against the idealization of certain insights from Marx's *Economic and Philosophical Manuscripts*. [German tr. *Prinzipien und Wege der Entwicklung der Psychologie*, Berlin GDR, 1963.]

[4] Ortega y Gasset believes that he rather than Heidegger should be credited with historical priority in conceiving of man as care: 'We come to define man as a being whose primary and decisive reality is his concern for his future . . . his pre-occupation. This is what human life is, first and foremost: preoccupation or, as my friend Heidegger put it thirteen years after me, *Sorge.*' See *La connaisance de l'homme au XX*e *siècle*, Neuchâtel, 1952, p. 134. The problem, however, is that neither he nor Heidegger recognized praxis as man's primary determination which implies authentic temporality. Care and the temporality of care are *derived* and *reified* forms of *praxis*.

[5] 'We are never at home, we are always beyond. Fear, desire, hope, project us toward the future and steal from us the feeling and consideration of what is.' *Complete Works of Montaigne*, Stanford, 1958, p. 8.

[6] Modern materialism was the first to eliminate the antinomy between the everyday and History, and to constitute a consistently monistic view of socio-human reality. Only materialist theory considers all activity as historical, and thus bridges the duality of the ahistorical everyday and the historicity of History.

[7] 'The mystery of the everyday ... turns out in the end to be the mystery of social reality in general. However, the dialectics immanent to this concept manifests itself in that the everyday both discloses and conceals this social reality.' G. Lehman, 'Das Subjekt der Alltäglichkeit', *Archiv für angewandte Soziologie*, Berlin, 1932–33, p. 37. The author incorrectly suggests that the 'ontology of the everyday' can be grasped through sociology and that philosophical concepts can simply be translated into sociological categories.

[8] Omitting or forgetting this subject expresses and creates one type of 'alienation of man'.

[9] Let us not forget that the terminology of existentialism is frequently an idealist–romantic, i.e. a concealing and dramatizing, transcription of revolutionary–materialist concepts. Finding this key permits a fruitful dialogue between Marxism and existentialism. I attempted to expose some aspects of Heidegger's subterranean and covert polemics with Marxism in my lecture 'Marxism and Existentialism', delivered in December 1962 in the Club of the Czechoslovak Writers' Union.

[10] Georg Büchner, 'Danton's Death', in *Plays of Georg Büchner*, London, 1971, p. 27.

[11] One of the less appreciated aspects of cybernetics is that it posed anew the question of what is specifically human, and in practice shifted the borderline between creative and non-creative human activity, between spheres that antiquity had defined as *scholé* and *ponos, otium* and *negotium.*

[12] The theory and practice of 'epic theatre' based on the principle of estrangement is only *one* artistic way of destroying the pseudoconcrete. Bertolt Brecht's connection with the intellectual atmosphere of the twenties and with the protest against alienation is obvious. One might also consider the work of Franz Kafka as an artistic destruction of the pseudoconcrete. See e.g. G. Anders, *Franz Kafka*, London, 1960, and W. Emrich, *Franz Kafka*, Frankfurt, 1957.

[13] Karl Marx, *Grundrisse*, New York, 1973, pp. 196–97. [Emphasis by Karel Kosík.–Tr.]

[14] See Karl Marx, *Theories of Surplus Value*, New York, 1952, p. 44.

[15] H. Freyer, *Die Bedeutung der Wirtschaft im philosophischen Denken des 19. Jahrhunderts*, Leipzig, 1921, p. 21.

[15a] Insufficient attention has been paid to the manner and the modifications in which the enlightened–materialist theory of interest has continued to live on into the twentieth century (for example, G. Anders translates Heidegger's *Sorge* as 'interest in the broadest sense'); similarly lacking is a complex analysis of the connections between Diderot's dialectics of the master and the servant and Hegel's dialectics of the master and the slave.

[16] In contrast to Shaftesbury who presumed immutable entities whose activity forms society, and for whom man is by nature a social being, i.e. is social even before society, Mandeville proved to be a true dialectician for whom opposites create something new, something that had not been contained in the premises.

[17] It would be very instructive to trace the history of the *concept* of the 'economic man'. The more fetishised science (political economy) becomes, the more does it view problems of reality merely as logical and methodological ones. Bourgeois political economy has lost the awareness of the connection between political economy's 'economic man' and the economic reality of capitalism which *reduces* man to the abstraction of the 'economic man' *really and practically*. It views 'homo oeconomicus' as a 'rational fiction' (Meyer), a 'necessary logical fiction' (H. Wolff), a 'working hypothesis' or as a 'useful caricature' (H. Guitton). On the other hand Gramsci (*Il materialismo storico*, pp. 266ff.) correctly emphasized the connection of the 'economic man' with the problems and reality of the economic structure that produces man's abstractness.

[18] 'The innumerable individual acts of circulation are at once brought together in their

characteristic social mass movement – the circulation between great functionally determined classes of society.' Karl Marx, *Capital*, vol. 2, p. 359.

[19] Social physics exists in an anti-metaphysical *illusion*: as a doctrine about man as an object and about his manipulability it can neither substitute metaphysics (philosophy) nor solve metaphysical (philosophical) problems.

[20] Transforming methodology into an ontology, or ontologizing empirical reality, is a frequent form of philosophical mystification. Every great epoch in philosophy destroys the reigning historical mystification. In his criticism of Aristotelian philosophy, Bacon criticised the ancients for having transformed a particular historical stage of developing human abilities, i.e. the lack of technology, into an ontology. See Paolo Rossi, *Philosophy, Science and Technology*, New York, 1970, p. 85.

Husserl described Galileo as at once a discovering and concealing genius, for having substituted as the founder of modern physics the idealized nature of natural sciences for reality (nature) itself. See Edmund Husserl, *Crisis of European Sciences and Transcendental Phenomenology*, Evanston, 1970, esp. par. 9.

[21] William Petty elaborated in *Verbum sapienti* (1665) a method to calculate the value of people in money; in 1736, Melon tried to prove that everything, including purely moral affairs, can be reduced to a calculation.

[22] Freyer, *op. cit.*, p. 17.

[23] J. Freyer, *Theorie des gegenwärtigen Zeitalter*, Stuttgart, 1955, p. 89.

[24] C. Wright Mills, *Sociological Imagination*, New York, 1959, p. 170.

[25] The weak point of idealist defenses of reason against existential interpretation is that they usually miss the connection of rationalist reason with a certain type of reality. Their arguments against existentialism are consequently hardly persuasive. See e.g. the illuminating polemic of Cassirer against Jaspers and others concerning the evaluation of Descartes, in E. Cassirer, *Die Philosophie im XVII. und XVIII. Jahrhundert*, Paris, 1959.

[26] The crucial question for Max Weber, who surrenders the individual's activity to irrationalism, is not the radical conflict between *Sein* and *Sollen* but the opinion that there exists no true, i.e. universal and necessary, knowledge of a value system. See Leo Strauss, *Natural Right and History*, Chicago, 1953, pp. 41–42.

[27] 'The individual who attempts to obtain these respective maxima is also said to act "rationally".' John von Neumann and Oskar Morgenstern, *Theory of Games and Economic Behavior*, Princeton, 1953, p. 9.

[28] Ch. Perelman and L. Tyteca, *Rhétorique et philosophie*, Paris, 1952, p. 112.

[29] Hegel was the first to analyse in depth this feature of modern times; see his *Glauben und Wissen*, Lasson, 1802, pp. 224, 225, 228, 229. Hegel's analysis of this specifically modern feature, of totality affected through a split, is dealt with in detail in Joachim Ritter, *Hegel und die französische Revolution*, Cologne, 1957, esp. pp. 32, 33.

[30] This split of consciousness is analysed in Husserl's important *Crisis of European Sciences* written on the eve of World War II. It might be in a way considered as the awakening of democratic awareness and a defense of reason against the danger of Fascism. Its philosophical content ranks among the seminal intellectual achievements of the first half of the twentieth century.

[31] Its historical limitations were, incidentally, exaggerated and abused by romantic reaction of all directions. It was natural that especially during World War II attempts would appear – especially in the bourgeois democratic camp – to rehabilitate Enlightenment and to defend reason against irrationalism. See e.g. Aron Gurwitch, 'On Contemporary Nihilism', in *Review of Politics*, 7, 1945, pp 170–198, particularly his defence of the eighteenth century against romantic irrationalist deformations, and also a lecture presented by A. Koyré in New York in 1944, at the 150th anniversary of Condorcet's death, in *Revue Métaphysique et de Morale*, 1944, pp. 166–189. Koyré

suggested that eighteenth century philosophy created a human ideal that remains the sole hope of mankind in its struggle against fascism.

[32] Perelman, Tyteca, op. cit., p. 122.

[33] Karl Mannheim, 'Das konservative Denken', *Archiv für Sozialwissenschaft*, 57, 1927, p. 492. Mannheim, burdened with sociologism and ignoring the real sources of modern dialectics, overestimated the role of irrationalism and romantism for the development of contemporory dialectical thought.

[34] Labriola describes factors as 'provisional concepts, which were and are a simple expression though not fully arrived at maturity'. They are 'the necessary product of a knowledge which is in the course of development and formation', and they 'arise in the mind as a sequence of the abstraction and generalization of the immediate aspects of the apparent movement'. A. Labriola, *Essays on the Materialist Conception of History*, New York, 1966, pp. 179, 145, 151. Similarly G. V. Plekhanov, *Development of the Monist View of History*, New York, 1972, pp. 13ff. *et passim.*

[35] G. V. Plekhanov, *Izbranniie sochineniia*, Moscow, 1956, vol. 2, p. 288. Did Professor Kareyev have students in the Czech lands as well?

[36] The materialist concept of the economic structure is inseparable from problems of labor and praxis, as we shall demonstrate in later chapters (especially in 'Art and its social equivalent' and 'Philosophy of labor'). Thus even the concept of 'economic structure' may degenerate into that of the 'economic factor' should this connection be absent.

[37] Valuable material relevant to this question is contained in the debate concerning the American constitution of 1787, in which representatives of different tendencies advocated their interests with candor unheard in the bourgeois society of later times. Hamilton: 'This inequality of property constituted the great and fundamental distinction in Society'. That same year, Madison wrote in the *Federalist* that 'the most common and durable source of factions has been the various and unequal distribution of property. Those who hold and those who are without property have ever formed distinct interests in society'. John Adams in a letter to Sullivan in 1776: 'Harrington has shown that power always follows property. This I believe to be as infallible a maxim in politics, as that action and reaction are equal, is in mechanics. Nay, I believe we may advance one step farther, and affirm that the balance of power in a society accompanies the balance of property in land'. See F. Coker, ed., *Democracy, Liberty and Property*, New York, 1947, pp. 73, 82, 120.

[38] This standpoint helps understand the unity of modern society and the structural interconnection of all its spheres, including economics (production for production, money–commodity–money), science (science as an absolute, i.e. an unlimited and ever-improving process of methodically acquiring and storing objective knowledge, a prerequisite for more complete control of nature), and of everyday life (accelerating tempo of life, absolute insatiability with pleasures, etc.).

[39] The vulgar pluralist standpoint is clearly manifest in opinions of John Dewey: 'The question is whether any one of the factors is so predominant that it is *the* causal force, so that other factors are secondary and derived effects'. 'Is there any one factor or phase of culture which is dominant, which tends to produce and regulate others, or are economics, morals, art, science, and so on only so many aspects of the interaction of a number of factors, each of which acts upon and is acted upon by the others?' J. Dewey, *Freedom and Culture*, New York, 1939, pp. 13, 16.

[40] It is a paradox of history, easy though it is to explain, that after World War I, bourgeois sociologists used Weber's theory of classes to argue the *impossibility* of a classless society (when it was necessary to prove the utopian character of the goals of the fledgling Soviet society), whereas after World War II the same theory provided arguments for the gradual end of classes and of class antagonisms, and for diminishing

class struggle in the most advanced monopolist-imperialist countries. For the first position, see Paul Mombert, 'Zum Wesen der soziale Klasse', in *Hauptprobleme der Soziologie*, 2, 1923, p. 267. For the second see J. Bernard and H. Schelsky, in *Transactions of the World Congress of Sociology*, 1956, vol. 3, pp. 26–31, and 1954, vol. 2, p. 360.

[41] 'Nostra namque, hoc est humana, sunt, quoniam ab hominibus effecta, quae cernuntur, omnes domos, omnia oppida, omnes urbes, omnia denique orbis terrarum aedificia, quae nimirum tanta et talia sunt, ut potius angelorum quam hominum opera, ob magnam quandam eorum excellentia, iure censeri debeant . . .' G. Manetti, *De dignitate et excellentia hominis*, Basel, 1532, pp. 129ff. Cf. also E. Garin, *Filosofi italiani del quattrocento*, Florence, 1942, pp. 238–42. Manetti (1396–1459) in his polemical ardor ignores that anything human can degenerate but this programmatic bias renders his trusting manifesto of humanism particularly charming. A hundred years later, Cervantes no longer shared *this* optimism, having arrived at a far more profound grasp of the problems of mankind.

[42] See e.g. the interpretation of romanticism and of unhappy consciousness in Jean Wahl, *Le malheur de conscience dans la philosophie de Hegel*, Paris, 1929.

[43] Attributes 'true', 'great', etc. should be pleonasms. Under certain circumstances, however, they provide the necessary clarification.

[44] These general observations could be graphically illustrated in the example of Picasso's *Guernica* which is of course neither an incomprehensible deformation of reality, nor a 'non-realistic' experiment in cubism.

[45] The formulation 'nothing but' has been encountered already in Chapter One, as a typical expression of reductionism.

[46] Marx described the reactionary apologetic character of bourgeois treatment of history and its concept of social reality in general in his apt comment, saying that it 'just consists in treating the historical conditions independent of activity', Marx, *The German Ideology*, New York, 1970, p. 60.

[47] 'The senses have their own history'. M. Lifshitz, *The Philosophy of Art of Karl Marx*, London, 1973, p. 78.

[48] The scientist who has no feeling for art is in the position of Janusz Kuczynski, and believes that the best textbook of political economy was actually written by Goethe under the attractive title of *Wahrheit und Dichtung*. See J. Kuzcynski, *Studien über schöne Literatur und politische Okonomie*, Berlin, 1954. Let us excuse the author by noting that his views were only 'an echo of the times'.

[49] The Plekhanovian method for writing a history of literature is reduced to the following procedure. First, a purely ideological history of the subject-matter is constructed (or, frequently, adopted ready-made from bourgeois scientific literature). Then an 'ordo et connexio rerum' is slipped under this 'ordo et connexio idearum', with the aid of frequently ingenious speculations. Plekhanov used to call this process 'the search for the social equivalent'. M. Lifshitz, *Voprosi isskusstva i literaturi* Moscow, 1935, p. 310.

[50] In this understanding of Marxism as a totality, Lenin agreed with Plekhanov, but even here he clearly viewed the concept of praxis entirely differently than Plekhanov had.

[51] Plekhanov, op. cit., vol. 2, p. 158.

[52] 'The difficulty is not in understanding that the Greek arts and epic are *bound up* with certain forms of social development. The difficulty is that they still afford us artistic pleasure and that in a certain respect they *count* as a norm and as an unattainable model'. Marx, *Grundrisse*, p. 111. [Emphasis Karel Kosík.–Tr.]

[53] Only in this light does the fragment in question clearly connect with other of Marx's works and opinions. Marx dealt with a similar problem when evaluating certain

classical political economists and when asking the question of objective truth in science. 'Every discipline of scholarship, including political economy and philosophy, has its own internal laws which guide its development, which are independent of the subjective caprices of individuals, and indeed are enforced even against subjective individual intentions or antipathies. On the case of Richard Jones, a successor of Malthus and an Anglican priest, Marx proved this *objective character of laws of science* which, when respected, lead to positive results, independent of the scientist's subjective intentions'. K. Kosík, *Dějiny filosofie jako filosofie: Filosofie v dějinách českého národa* [History of Philosophy as Philosophy: Philosophy in Czech History], Prague, 1958, p. 15.

[54] See R. Ingarden, *The Literary Work of Art*, Evanston, 1973; also V. Vinogradov, *Problema avtorstva i teoria st'ilei* [The Problem of Authorship and the Theory of Style], Moscow, 1961, p. 197; L. Doležel, *O stylu moderní, české prózy* [The Style of Modern Czech Prose], Prague, 1960, p. 183.

[55] A false method again ends up making inadvertent substitutions that the scholar overlooks: he discusses 'reality' whereas his false method has meanwhile transformed reality into something else and has reduced it to 'conditions'.

[56] A. Hauser, *The Philosophy of Art History*, New York, 1959, pp. 185f.

[57] Since theoretical thinking does not disappear with the conditions that gave rise to it either, the seventeenth century discoveries concerning human nature are valid in this century, too. Every theory of history and of social reality therefore falls back on Vico's seminal discovery of the historical character of human nature. 'Human nature is entirely historicised, it is a nature in making. It is no longer a permanent nature that could be known outside its historical manifestations. It forms a one with these manifestations which constitute moments of its present as well as its future'. A. Pons, 'Nature et histoire chez Vico', *Les études philosophiques*, Paris, 1961, No. 1, p. 46. Marx's high regard for Vico is generally known.

[58] It is often overlooked that Hegel's logical apriorism with which he considers history as the application of Spirit in time and thus as applied logic, as the unfolding in time of moments of the Spirit, otherwise essentially timeless, is the most grandiose idealist attempt of modern times to overcome or turn back relativism and historicism.

[59] In modernizing Hegel's concept of reality as 'Bedeutung, Weltbedeutung, Kulturbedeutung', Emil Lask is clearly viewing Hegel as an orthodox Kantian and a disciple of Rickert. Cf. Lask, *Schriften*, vol. 1, p. 338.

[60] The primitivism and relativism of closed-horizon theories are opposed, as expressions of twentieth-century antirational theories, by Th. Litt, in *Von der Sendung der Philosophie*, Wiesbaden, 1946, pp. 20f, who calls for philosophy to be the search for universal truth. The idealism of this critique of antihumanism is in its failure to recognize not only knowledge, but also praxis as a crucial way of overcoming relativism.

[61] 'The great discovery of the eighteenth century is the phenomenon of memory. By remembering, man escapes the purely momentary; he escapes the nothingness that lies in wait for him between moments of existence.' G. Poulet, *Studies in Human Time*, Baltimore, 1961, pp. 23f. The author documents his view with references to works of Quesnay, Diderot, Buffon and Rousseau.

[62] It follows from what has been stated previously that this 'life' includes the possibility of many interpretations, every one of which adopts different aspects of the work.

[63] The connection between categories of rejuvenation and reproduction in Hegel's and Marx's philosophy has been correctly pointed out by M. Lifshitz (*Philosophy of Art of Karl Marx*, pp. 109ff.). 'The rejuvenation of the spirit is more than a return to the former self; it is a self-purification, a working over.' Hegel, *Philosophy of History*,

vol. 1, p. 11. The great thoughts of Novalis, which are scattered throughout the Christian-romantic ether of his philosophy, identify totalization with animation. Cf. Th. Hoering, *Novalis als Philosoph,* Stuttgart, 1954, p. 45. Hoering's extensive but poorly organized work suffers from one basic shortcoming, in that it dilutes the specific contributions of Novalis' thinking in the general dialectical atmosphere of his times; subjected to such treatment, Novalis emerges as a junior Hegel.

Translator's Note: Throughout this section Kosík is opposing two Czech terms which parallel Heidegger's opposition of *Sorge* and *Besorgen* as the 'ontological' and 'ontical' aspects of Dasein's involvement in the world. These German terms are rendered 'Care' and 'Concern' in the English translation (*Being and Time*, trans. J. Macquarrie and E. Robinson, New York, 1962). Since the translation of *Besorgen* as 'concern' obscures many of the 'economic' senses of the German (and corresponding Czech) term, the term will here be rendered 'procuring' – which preserves both the economic connotations and etymological relation to 'concern', through the Latin *Cura*.

CHAPTER III

PHILOSOPHY AND ECONOMY

PROBLEMS OF MARX'S *CAPITAL*

Interpretation of the text

The reader who has to plough through *Capital* several times in order to comprehend its specialized economic sense and to get the clear meaning of concepts such as value, falling rate of profit, surplus value, the processes of producing capital and surplus value, etc., does not usually ask about the overall meaning of Marx's work. The question either never enters his mind, or he is satisfied with answering it with some general considerations in which comprehending the text never becomes a *problem*. In addition, because Marx's text is a difficult work, the average reader studies it as presented in a political economy *textbook*, designed to popularize the complex subject-matter. However, what are the difficult passages of the text, what passages are seen that way, and what does a popularization entail? First of all, Marx's *extensive* text is abridged. Second, all that would interfere with elaborating narrow problems of economics is routinely culled out of the text. Analyses of obsolete nineteenth-century data are deleted or replaced by more recent data. *Similarly*, passages that from the 'strictly scholarly perspective' seem to be no more than speculations or perhaps dispensable philosophical contemplation not directly connected with economic problems are left out as well. Since the textbook is a *guide* to studying the text, the reader follows it in attributing greater or lesser importance to different sections of *Capital. This* way of reading the text, however, imports problems of which the reader — and frequently even the textbook authors — are unaware. *This* is not a reading of the text written by Karl Marx but of a different, altered text. Popularization which had at first appeared as *merely* rendering the text more accessible, turns out to be a particular *interpretation* of it. Every aid to understanding a text has limits beyond which it ceases to fulfil its auxiliary introductory and clarifying role, and assumes the opposite role of obscuring and distorting. A popularization which is unaware of its own limits and does not see itself critically, as merely one *particular* interpretation of the text, as an interpretation which for didactic purposes considers only certain *particular* aspects of the text and *consciously* omits others, ends up *unconsciously* engaged in an entirely different activity: instead of interpreting the text it modifies it and uncritically invests it with a *different* sense.

But why does the text have to be interpreted at all? Does it not speak for itself, and in a clear enough language? Who could have expressed the intended thought more clearly and poignantly than the author himself? What does it mean to say that the author imparted a certain meaning to the text? (We mean 'text' in the broad sense, i.e. not only literature but also paintings or sculptures, any structure of significations.) From what can we judge the author's subjective intentions? For the majority of extant texts we rely only on the text itself. We do not always know enough about the author's subjective intentions. Even when such information is available, it hardly solves the problem, for the relation of the text itself to reports about its author's intentions is not an unambiguous one: such evidence might help explicate the meanings of the text, but these can in principle be captured even without it. Compared with the text (the work) itself, 'documents' play a complementary and a secondary role. The text may even say something other than the testimony does: more, or perhaps less. The author might not have fulfilled his intention, or he might have exceeded it, in which case the text (the work) contains 'more' than he had anticipated. As a rule, the intention is congruent with the text, and is thus *expressed* in and through the text: only the message of the text testifies to the author's intentions. The text is the starting point for its interpretation. The interpretation starts out from the text in order to return to it, i.e. to explain it. If it does not return, the familiar inadvertent substitution of one task for another will take place and instead of being interpreted, the text will end up being examined as a testimony to its times and conditions.

The history of a text is in a certain sense the history of its interpretations: every period and every generation emphasizes different aspects of it, attributes greater importance to some than to others, and accordingly reveals different meanings in the text. Different times, generations, social groups and individuals can be blind to certain aspects (values) of the text and will find them meaningless, concentrating instead on aspects which in turn appear unimportant to their successors. The life of a text is thus a process of *attributing* meanings to it. Does this attribution concretize meanings which the work *objectively* contains, or does it import new ones? Does there indeed exist *the* objective meaning of the work (the text), or can the work be grasped only through different subjective approaches? It seems that we are locked in a vicious circle. Is it possible to interpret the work authentically, in a way that would capture its objective meaning? If it were not possible, then any attempt at an interpretation would be senseless, since the text could be grasped exclusively in subjective approaches. However, if an authentic interpretation is possible, how should one square this with the fact that every text is interpreted differently and that the history of a text is the history of its various interpretations?

Interpreting a text assumes that a substantiated interpretation of it can be distinguished in principle from textual distortions or modifications. We require the following of an interpretation:

That it leave no opaque, unexplained or 'accidental' passages in the text.

That it explain the text both in its parts and as a whole, i.e. that it deal both with its individual sections and with the structure of the work.

That it be complete, and not suffer from internal contradictions, questionable logic or inconsistencies.

That it preserve and capture the *specificity* of the text and incorporate this specificity as a constitutive element of the structure and the comprehension of the text.

If it is possible to arrive at an *authentic* understanding of a text, and if *every* interpretation is an historical form of the text's existence, then the authentic interpretation will indispensably include a critique of all previous ones. Partial or one-sided interpretations will then appear as layers that have sedimented on the text over the years, as *historical* forms of the text's existence (the text itself being always distinct and independent of them), or else as manifestations of various concepts that have guided the interpretation: concepts of philosophy, science, art, reality, etc. *Every* interpretation of a text is *always also* its *evaluation*, be it unintended and thus unsubstantiated, or conscious and reasoned: glossing over certain parts or sentences which are felt as unimportant (different ones at different times), or simply misunderstanding certain passages (depending on the age, education, cultural background of the reader) and subsequently 'neutralizing' them in itself amounts to an implicit evaluation, inasmuch as it distinguishes in the text between the significant and the less significant, between the relevant and the obsolete, the important and the secondary. The history of interpretations of Karl Marx's *Capital* shows that every interpretation covers up a particular concept of philosophy, of science and reality, of the relation between philosophy and economy, etc., which informs both the explication of individual concepts and thoughts, and the construction of the work as a whole.

A number of expositions of *Capital* have violated the first rule of interpretation: that an exposition, to be authentic, should leave no 'opaque' and unexplained places in the text. The exposition should not divide the text into one part which can be explained by a certain principle and another which does not lend itself to this interpretation, and from the perspective of this principle is thus mute and unimportant. Since many expositions of *Capital* have failed to cope with its 'philosophical passages', and considered the philosophical problems of *Capital* to be a dispensable factor (if they indeed discovered these problems at all, other than in some explicit passages which from the point of view of economic issues appeared irrelevant anyway), this violation of the formal rule of interpretation has posed a major obstacle even to understanding the *character* of the text. All such interpretations broke the single text down into two, dealt with one according to a particular principle and found the other inexplicable. That one then became incomprehensible and insignificant.

We take an interpretation to be authentic if the *specificity* of the text is a constitutive element of the principle of its exposition, as it unfolds in the course of this exposition. The interpretation substantiates the text's specificity. The text can of course fulfil functions in which its specificity plays no role. Shakespeare's historical dramas can be considered and employed as a testimony to his time. K. H. Mácha's poem *May* can be studied from the perspective of the author's biography. The history of ideology may include dramas, poems, novels and stories. It will abstract from the specificity of their genre and examine them exclusively as manifestations of different world outlooks. Common to these approaches is that they all erase or ignore the specificity of the works as lyrical poetry, as a novel, tragedy, epic poem, etc. The specificity of the text is not an abstract universal framework, a categorization of genre, but a specific principle of the work's *construction*. This specificity is the result of investigation, and is not known at its outset. Consequently, it does not regurgitate trivialities or impute abstract principles to the text as much as it seeks for what is specific in it.

There never was much dispute about the *Wealth of Nations, Principles of Taxation, General Theory of Employment*, etc. being works of economics, and specifically so. However, *Capital* has from the very beginning been provoking the uneasiness of a number of interpreters who would agree on one thing only: it is not a work of economics in the usual sense of the word, and it conceives of economics in a peculiar way, splicing it with sociology, philosophy of history, and philosophy. Judging by the history of the interpretations of *Capital*, the relation between science (economics) and philosophy (dialectics) appears to be its key problem. The relationship of economics and philosophy is not just another partial aspect of Marx's work (and useful research has been done on his use of statistics, on the incorporation of historical material, on the use of fiction in *Capital*, etc.) Rather, it provides access to the very *essence* and *specificity* of *Capital*.

Different interpretations of *Capital* have attempted to uncouple its science from its philosophy in several ways. They all in some way divorce science from philosophy, specialized scientific investigation from philosophical assumptions, and thus lead via different paths to one result: to a science and a philosophy that are mutually indifferent.

In one instance, science (economics) and philosophy *both* end up as superfluous. This interpretation translates economic movement into logical movement, and transcribes Marx's *Capital* so as to render scientific observations in the language of philosophy. The economic content is irrelevant to and independent of the logical categories. This conception considers Marx's work first and foremost as an *applied* logic which uses economics to demonstrate its own movement. The economic movement is entirely external to philosophy because it is only an agent of the movement of logic. The truth of economics is expressed in the movement of logic. It is quite foreign to and independent of the economic content, because the movement of logic could just as well have been expressed through any other specialized scientific discipline. Philosophy related to economics is conceived also as a mere methodological—logical substrate or as applied logic. The task of the interpreter is to decant from this applied logic a pure logic, and behind the movement of such categories as declining rate of profit, transformation of surplus value into profit, price formation, etc., to discover and distill pure logical categories of movement, contradiction, self-development, mediation, etc. But we could similarly consider *Capital* to be an applied grammar: its economic content is formulated according to certain rules of linguistics which also could be abstracted from the text. Inasmuch as the connection between science and philosophy is seen in the stratification of the text (the text having an economic as well as a logical—methodological meaning), there is no difference between Marx's *Capital* and F. Palacký's *History of the Czech Nation*: the text of Palacký can be considered an applied logic just as Marx's can. If this is the task the interpreter has set about fulfilling, he has to conclude by answering one question: Why did Marx write a book of economics and Palacký one of history, and why did neither of them write a 'pure logic' rather than an 'applied logic'? If an interpretation considers an economic or an historical text to be an 'applied logic' from which a 'pure logic is to be distilled, he must crown this exacting labor by the most important task: by proving that the logical and methodological categories he had used for analysing a specific economic or historical reality are valid generally, and that they are applicable even beyond the framework of the reality in question. Interpretations with a logical or a methodological bent do not try to critically examine the economic content of *Capital* and they do not even try to further develop and elaborate its economic problems. Ready-made results of economic analyses are with no further inquiry automatically taken as correct, and the interpretation traces only the logical and methodological path that had led to results whose fundamental validity is not questioned.

Another interpretation defends the validity of *Capital's* economic content against modern bourgeois critiques but concludes that this economic content lacks a proper philosophical rationale. This can apparently be furnished by phenomenology.[1] *Capital* thus turns out to be a valid economic analysis without a proper philsophical foundation. However, complemented with the necessary philosophy, the sense of the text would change, and a Marxist political economy would turn into an extensive phenomenology of objects. A materialist analysis of capitalist economy would turn into a phenomenological description of a world of things.

A third interpretation of *Capital* asks the question: 'Is this pure political economy, an analysis of mechanisms, or rather an existential analysis of economics, with a metaphysical and transeconomic significance?'[2] The question posed in this way is essentially a result of half-truths. Is *Capital* indeed pure political economy, a theory of mechanisms, i.e. a science *scientistically* conceived? Since this interpretation does not consider Marxist political economy to be a science in *this* sense, it concludes that Marx is no economist in the real sense of the word.[3] Since Marxism is neither a scientistic—empirical kind of science, nor an instance of vulgar economics, it is no science at all. What is it then? Marxist political economy is apparently an existential philosophy that considers economic categories as mere signs or symptoms of a concealed essence, of the existential situation of man.[4]

Conversely, a fourth interpretation stresses the necessity to separate the positive detailed economic part of Marx's work from philosophical speculations (dialectics). It recognizes in Marx a great economist who however has to be defended from Marx the philosopher. Marx's economic analyses are based on a scientific economic method which is not only at variance with dialectics but is indeed entirely independent of it, so that the scientific value of Marx's analyses is preserved *despite* the metaphysical—speculative dead wood heaped over it.[5] In this setting we are hardly interested in indulgent assurances that Marx was a true scientific talent, which sounded naive and grotesque already around World War I. We are more intrigued by the sense and the content imputed to the term 'science'. This interpretation radically separates science from philosophy because its concept of science is based on the image of an empirical model: one of presuppositionless observation and analysis of facts, which is of course a mere prejudice belied in daily practical life.[6]

To Abolish Philosophy*?

Let us pursue the question from another aspect. Can the relation between philosophy and economics in *Capital* be clarified by analysing Marx's intellectual development? We are not nearly as much interested in describing in detail his intellectual history as in tracing its inner logic. Since, however, the logic of the thing differs from subjectivist constructions or ideas about the logic of the thing, we must formulate it as the *result* of critically investigating *empirical* material, which is the starting point as well as the goal of this investigation: the investigation can claim to be critical and scientific only insofar as it gathers all possible empirical material, and as long as the 'inner logic' it discovers captures this totality completely and concretely, i.e. as long as it gives it an objective meaning and explains it. The objective meaning and the internal problems of the text are revealed through its interpretation in the 'intellectual milieu' and the socio-historical reality. The intellectual development of a thinker or of an artist therefore cannot be investigated by thoughtlessly narrating his lifestory or by unproblematically 'commenting' on his works or opinions.

We are interested in the question of whether the relationship of philosophy and economy (science) changed in the course of Marx's intellectual development, and in the way that Marx conceived of and formulated this relationship in different phases of his own development. This question has been the center of attention of Marxists and Marxologists for many years now, in the familiar arguments over the 'young Marx'. This discussion has not led to overwhelming results. Instead of concrete investigations, students have frequently presented only general methodological precepts, and their 'commentaries', usually unencumbered by heeding their

---

*Where German philosophy (and the German translation of Kosík) uses the term *aufheben* Kosík employs the Czech *zrušit*. The dual meaning of *aufheben* is lost in the Czech translation, as in most ways of rendering it into English. *Zrušit* literally means 'cancel' or 'abolish.'—Tr.

own methodological advice, are extraordinarily sterile. If, as generally asserted, the anatomy of man is indeed the key to the anatomy of apes, and if the work of the young Marx has indeed to be comprehended from the work of the mature Marx and from evolving revolutionary materialism, then one might expect that advocates of this rule will also stick to it and consequently present an interpretation of the *Manuscripts* based on an *analysis* of *Capital*. In reality, however, the *Manuscripts* are interpreted in isolation from Marx's total development (which is one of the reasons for the repetitiveness, ennui and superficiality of dozens of essays entitled 'The Young Marx'), and the explication of the problematique is based on one covert assumption: on a muddled idea about the dynamics of Marx's intellectual development. This muddle, amounting to a lack of critical attitude, is the graveyard of science and of scientific explication, because it allows the investigation to move with naive confidence over terrain that is through-and-through problematic. Uncritical naiveté has not the slightest idea that specific conceptual means are needed to grasp someone's intellectual development. Without them, the empirical material is either incomprehensible or elusive, or it is senseless and conceals its own 'hidden truth'.

A sufficient number of 'case studies' will permit the construction of several basic models or schemes of the dynamics of intellectual development. These models have two functions: first, they are an *intuitive representation* of intellectual development and of its *dynamics as a whole* (of its direction, curve, regressions, complexities, deviations); and second, they provide a *conceptual* means for *comprehending* individual works, periods, partial opinions. Without claims to completeness and exhaustive characterization, we suggest that the majority of 'cases' will fall under one of the following basic models of intellectual development dynamics:

(1) *The model of empirical-evolutionary development* in which a particular elementary basis of a world view flourishes and, influenced by and reacting to events, grows more profound and more universal. It rids itself of outdated or incorrect elements and substitutes adequate ones for them.

(2) *The model of evolutive development through crises*, marked with sharply separated periods signifying abrupt shifts from one concept of the world to another, a conversion from one 'profession' to another, in which the past or the preceding period is negated as one-sided, as an error or a delusion.

(3) *The model of holistic–concretizing development* in which a rich world view is formulated in an early stage of creative reasoning. Its *basic* motives and problems are never abandoned or transcended but are rather

developed further, rendered more precisely, and formulated more exactly on basis of subsequent study and praxis.

The unconscious and unanalysed scheme of most interpretations of Marx's intellectual development assumes that the transition from the *Manuscripts* to *Capital* is a transition from philosophy to science. Whether this development is estimated positively or negatively, as progress or degeneration,[7] it is always characterized by gradually abandoning philosophy and its problematique for science and the exact scientific problematique.[8] Marx's intellectual development epitomizes and realizes the radical demand of left Hegelians: *to abolish philosophy*.

How might philosophy be abolished and how has it been abolished in Marx's work?

Philosophy can be abolished by realizing it.

Philosophy can be abolished by turning it into a dialectical theory of society.

Philosophy can be abolished when it falls apart and survives as a residual science: as formal or dialectical logic.

Philosophy can be abolished by realizing it. This statement amounts to an *idealistic* formulation of the relationship between philosophy and reality: society with all its contradictions finds an appropriate historical expression in philosophy, and the philosophical expression of real contradictions becomes the ideological form of praxis that solves them. Philosophy plays two roles in its relation to society: the epoch, the society or the class develop their own *self-consciousness* through philosophy and its categories. At the same time, they find in philosophy and in its categories categorial forms of their own historical praxis. Philosophy is not 'realized' but rather, reality is 'philosophised'. That is to say, reality finds in philosophy both an historical form of self-consciousness and an ideological form of praxis, of its own practical movement and problem-solving. Those who would 'abolish philosophy by realizing it' see social movement as traversing the movement of human consciousness where it develops categorial forms of its own realization. Apart from that, 'realization of philosophy' is an inverted expression for realizing the latent *possibilities* contained in reality.

An idealist conception stands these relations on their head and inverts the relationship between the original (the reality) and the 'still picture' (philosophy). It conceives of reality as of a realized or non-realized philosophy. Since the original is superior to the reproduction, the truth of reality has to be grasped *as derived from* philosophy (the original). The radical statement of abolishing philosophy by realizing it expresses neither the truth of philosophy nor that of reality, but merely the contradictory character of utopism that seeks to realize a *pale reflection* of reality.[9] Since philosophy is the reality of the epoch concentrated in *thoughts*, philosophical self-consciousness may fall for the self-delusion that reality is a reflection of philosophy and that its relation to philosophy is that of something which will or should be realized. In this idealistic perspective, philosophy turns into unrealized reality. Philosophy, however, is supposed to be more than realized. It is supposed to be abolished through realization, no less, since its very existence is an expression of unreasonable reality. To abolish alienation means: to abolish the existing unreasonable society as a realization of philosophy, and at the same time to abolish philosophy by realizing it, for its very existence testifies to the unreason of reality.[10]

Considered from this vantage point, the slogan of abolishing philosophy by realizing it is nothing but an eschatological fiction. First of all it is not true that philosophy is merely an alienated expression of alienated conditions and that this description exhausts its character and mission. Only particular historical instances of philosophy might amount to false consciousness in the absolute sense, but from the perspective of philosophy in the real sense of the word, these would *not* amount to philosophy. They would be mere systematizations and doctrinaire interpretations of biases and opinions of the time, i.e. ideologies. The suggestion that philosophy is *necessarily* an alienated expression of an inverted world, because it has always been a class philosophy, might have originated from a *misreading* of the *Communist Manifesto*. This suggestion would have the text read thus: 'History of mankind does not exist, there is only the history of class struggle', instead of the actual text, 'the history of all hitherto existing society *is* the history of class struggles'. Then it would follow that every philosophy has been *exclusively* a class philosophy. In reality, however, that which has a class character and that which has a human character have formed a dialectical unity throughout history: every historical epoch of *mankind* was spearheaded and represented by a particular *class*, and mankind and humanism have been filled with a concrete historical content which is both their concretization and their historical limit. The historicity of conditions is substituted here again for the historism of reality, and philosophy is vulgarly conceived as a manifestation of conditions, rather than as the truth of reality.

The slogan of realizing philosophy has many meanings. How can one recognize whether what is being realized is indeed philosophy and only philosophy, or whether it is something else, something that perhaps goes beyond philosophy or does not measure up to it? And indeed, even if what is being realized were philosophy, is it realized entirely and with no leftovers, and is then reality an absolute identity of consciousness and being? Or do perhaps some ideas of philosophy 'reach beyond' reality, and subsequently lead philosophy into a conflict with reality? What does it mean that the bourgeois society is the realization of the reason of Enlightenment? Is the totality of bourgeois philosophy identical with the totality of bourgeois society? And if a bourgeois society amounts to incarnated philosophy of the bourgeois epoch, will the demise of the capitalist world lead to the extinction of this philosophy? Who is to judge, and who will judge in the future, whether indeed *reason* has been realized through abolishing philosophy and whether society is indeed reasonable? Which level of human consciousness will recognize whether reality has not merely been rationalized and whether reason is not again being realized in the *form* of unreason?

All these unclear points stem from a profound contradiction in the very conception of reason and reality, one that is shared by all eschatological reasoning: history exists up to a point, but it ends at a critical moment. A dynamic *terminology* conceals a *static* content; reason is historical and dialectical *only up to a certain phase* in history, up to a turning point, whereafter it changes into trans-historical and non-dialectical reason.

The eschatological formulation of abolishing philosophy through realizing it obscures the *real problem* of modern times: does man *still* need philosophy? Have the position and the mission of philosophy in society changed? What role does philosophy play? Is its character changing? Naturally, these questions do not affect the empirical facts that philosophy is still extant, that it is practiced, that books on philosophical topics are being written, and that it is a specialized discipline and profession. The question is elsewhere: does philosophy continue to be a special form of consciousness, *indispensable* for grasping the *truth* of the world and for arriving at a *truthful* comprehension of man's position in the world? Does truth still happen in philosophy, and is philosophy still considered a sphere in which opinion is distinguished from truth? Or has philosophy taken over from mythology and religion as the universal mystifier, the spiritual medium necessary for mystification? But perhaps it has been denied even this honor, what with modern technology having provided the mass media, even more efficient means of mystification. Does then the continued existence of philosophy prove that the realization of reason, so frequently heralded, has after all *not* occurred yet? Or does the periodic alternation of chiliasm and skeptical sobering up, and the permanent disharmony between reason and reality, perhaps indicate that reason and reality are indeed dialectical and that their called-for absolute identity would amount to the *abolition* of dialectics?

A different way of abolishing philosophy is to transform it into a 'dialectical theory of society' or to dissolve it in social science. This form of abolishing philosophy can be traced in two historical phases: the first time during the genesis of Marxism when Marx, compared with Hegel, is shown to be a 'liquidator' of philosophy and the founder of a dialectical theory of society,[11] and the second time in the development of Marx's teachings which his disciples conceive of as social science or sociology.[12]

The genesis of Marxism is interpreted against the background of the *dissolution* of Hegel's system as the culminating phase of bourgeois ideology. The synthesis and totality of Hegel's philosophy had disintegrated into elements. These were in turn absolutized, and they formed bases for *new* theories: for Marxism or existentialism. Historical research has correctly pointed out[13] that the disintegration of Hegel's system resulted in no intellectual vacuum; the very term 'disintegration' conceals and masks a wealth of philosophical activity which gave rise to the two important philosophical orientations of Marxism and existentialism. The shortcoming of these observations is that they consider Hegel the pinnacle and synthesis, compared with whom Marx and Kierkegaard necessarily appear one-sided. This opinion is inconsistent. Abstractly speaking, one could advocate any one of the three philosophical standpoints, consider it the absolute, and criticise from its vantage point the other two as the incarnations of one-sidedness. From the absolute standpoint of Hegel's system, the subsequent development would appear as the collapse of total truth, and the different orientations that emerged from it as emancipated elements of that collapse. From the viewpoint of Kierkegaard, Hegel's philosophy would be a lifeless system of categories with no room for the individual and his existence. Hegel might have constructed palaces for ideas but he kept people in shacks. Socialism is the continuation of Hegelianism.[14] Marxism criticises Hegelianism and existentialism as so many varieties of idealism: objective and subjective. However, where is the objective measure for the 'absoluteness' of one's own standpoint? Under what circumstances does an opinion become the truth? Opinion becomes truth if it demonstrates and proves the truth of its opinion. This includes demonstrating its own ability to comprehend through philosophical activity and reasoning the other standpoints as well, to explain their historical justification, as well as the historical conditions for transcending them, to *realize the truth* of the criticised standpoints and *thereby* to prove their biases, limitations and falsity. However, the truth of this proof is historical. It is constituted forever anew and it *proves* again and again its true character. The historical development of this truth will consequently also include periods in which 'absolute truth' or the truth of the 'absolute standpoint' actually *collapses* into elements which it had historically transcended and integrated in itself. Materialist philosophy can in certain historical periods disintegrate into the philosophy of the 'Absolute Spirit' (Hegelianism), whose critical complement is the philosophy of existence and moralism. This, too, is an indirect proof that Hegel and Kierkegaard can be comprehended on the basis of Marx, but not vice versa.

One argument for dialectically abolishing philosophy in social science is the statement that the materialist inversion of Hegel is not a transition from one philosophical position to another, that it is not a continuation of philosophy. This statement is extremely inaccurate, since it obscures the specificity of the 'transition' from Hegel to Marx. From the standpoint of materialist dialectics, neither the history of philosophy as a whole nor its individual stages can ever be interpreted as a 'transition from one philosophical position to another', because such an interpretation presupposes an immanent evolution of ideas, which materialism denies. Inasmuch as the development from Hegel to Marx *is not* a transition from one philosophical position to another, it does not in any way imply the need to 'abolish philosophy', just as the development from Descartes to Hegel did not abolish it, though it was not (*merely*) a transition from one philosophical position to another. Equally confusing is the second argument, according to which 'all the philosophical concepts of Marxian theory are social and economic categories, whereas Hegel's social and economic categories are all philosophical concepts'.[15] Here, too, the general is presented as the particular and its specificity is obfuscated. Marxist critique detects a social and economic content in *every* philosophy, including the most abstract, because the subject who elaborates a philosophy is no abstract 'spirit' but a concrete historical person whose reasoning reflects the totality of reality, complete with his own social position. Every concept contains this 'socio—economic content' as its moment of *relativity*, which is both a degree of approximation and imprecision, *and* the capacity to *improve* human cognition and to make it more precise. Inasmuch as every concept contains a moment of relativity, *every* concept is both an *historical* stage of human cognition and a moment of *improving* it. The theory of 'abolishing philosophy', however, grasps the 'socio—economic content' of concepts subjectively. The transition from philosophy to a dialectical social theory not only realizes the transition from philosophy to non-philosophy, but above all *reverses* the meaning and the sense of concepts that *philosophy* had discovered. The statement that all philosophical concepts of Marx's theory are socio—economic categories expresses the *double metamorphosis* Marxism has undergone in transition from philosophy to social theory. First, the historical reality of *discovering* the character of economics is obscured. Second, man is imprisoned in his subjectivity: for if all concepts are in essence socio-economic categories, and express *only* the social being of man, then they turn into forms of man's self-expression, and every form of objectivation is only a variety of reification.

Abolishing philosophy in dialectical social theory transforms the *significance* of the seminal 19th century discovery into its very opposite: praxis ceases to be the sphere of humanizing man, the process of forming a socio—human reality as well as man's *openness* toward being and toward the truth of objects; it turns into a closedness: socialness is a cave in which man is walled in. Images, ideas and concepts that man takes for spiritual reproductions of nature, of material processes and of objects existing independently of his consciousness, are in 'reality' a social projection, an expression of man's social position in the *form* of science or of objectivity. In other words, they are *false* images. Man is *walled in* in his socialness.[16] Praxis which in Marx's philosophy had made possible both objectivation and objective cognition, and man's openness toward being, turns into social subjectivity and closedness: man is a prisoner of socialness.[17]

The Construction of Capital

The opening paragraph of *Capital* reads: 'The wealth of those societies in which the capitalist mode of production prevails, presents itself as "an immense accumulation of commodities", its unit being a single commodity. Our investigation must therefore start with the analysis of a 'commodity'. The concluding section of the entire work, the unfinished fifty-second chapter of the third book, is devoted to an analysis of classes. What connection is there between the beginning and the end of *Capital*, between its analysis of *commodities* and its analysis of *classes*?

The very question raises suspicions and doubts. Is not an attempt to disguise in the garb of a smart and heavy question the trivial fact that every work has a beginning and an end? Is this questioning not a cover-up for the most arbitrary licence with which someone randomly perused the book's beginning and end, and now pretends to have made a 'scientific' discovery in juxtaposing them? What would science come to if it were to search for 'internal connections' between opening and closing sentences? Such scepticism could be further strengthened by noting that the third volume of *Capital* was published post-humously and that its closing chapter remains a fragment. It is indeed possible that the entire fifty-second chapter is only an *accidental* conclusion, and that the entire suggestion of a more 'profound' connection between the beginning and the end of the work, between commodities and classes, therefore stands on quicksand.

We do not intend to examine the extent to which Engels' editing of *Capital's* third volume corresponds in every detail to Marx's intentions, and whether Marx would indeed have concluded his work with a chapter on classes. Speculations and hypotheses of this kind are all the less pertinent since we see the connection between the opening and the conclusion of *Capital* not merely as a catenation of the first and the last sentences, but as an immanent structure and principle of the work's construction.

We can thus formulate the original question more precisely: what is the relation between the immanent structure of *Capital* and its external organization? What is the connection between the *principle of its construction* and its literary form? Are its analyses of commodities and classes only the starting and closing points of the external organization of the subject-matter, or does their connection reveal the *structure* of the work? Though these *particular* questions have so far not been posed in the literature, the problematique they touch on is not new. It has appeared, for example, in expositions of the points shared by *Capital* and Hegel's *Logic*, or in well-known aphorisms that one cannot fully comprehend *Capital* without having studied and comprehended the *whole* of Hegel's *Logic*, and that though Marx did not leave behind a *Logic*, he did leave the logic of *Capital*.[18] These problems are also contained in the suggestion that *Capital* is both Marx's *Logic* and his *Phenomenology*.[19] Finally, it transpires in the somewhat artificially construed argrument over why Marx revised the original plan of *Capital*, 'in 1863', and substituted a new one for it, which is supposed to be the basis for the final version.[20]

At any rate, the carefully thought-out *architecture* and the minutely designed internal construction of the work are striking and prominent features of *Capital*. Marx himself saw merit in that his work formed 'an artistic whole' (*ein artistisches Ganzes*). One might infer that the structure of *Capital* is an 'artistic' affair concerning the literary treatment of the subject-matter. One might say that the author had mastered the subject-matter scientifically, and then selected the *form* of an 'artistic whole' or of a 'dialectical organization' for its literary shape. Changes of plans could then be easily explained as stages in the literary shaping of the subject-matter which had been scientifically mastered and analysed previously. But even when Marx discusses *Capital* as an 'artistic whole', he emphasizes the difference between his own dialectical *method* and the analytical—comparative procedures of Jacob Grimm.[21] The architecture of *Capital* as an 'artistic whole' or as a 'dialectical organization' thus has to do *both* with the literary treatment of the subject-matter *and* with the method of its scientific exposition. At this point, interpretations usually halt. Here they have struck pay dirt and can fruitfully investigate the logical structure of *Capital*, comparisons of identity and difference between Marx's logical concepts and those of Hegel, or undertake the even more challenging task of abstracting from *Capital* an entire *system* of categories of dialectical logic.

But *Capital* is a work of economics and its *logical* structure must therefore match in some way the structure of the analysed *reality*. The structure of *Capital* is not a structure of logical categories to which the reality under investigation and the treatment of it are to be subordinated. Rather, a scientifically analysed *reality* is adequately expressed in a 'dialectical organization'. It is executed and realized in a particular *corresponding* logical structure.

The peculiar character of *reality* is the cornerstone of the structure of *Capital* as a 'dialectical organization', from which it can be comprehended and explained. The literary treatment in the 'form' of an artistic whole, the dialectical method of 'unfolding', and the revealing of the specific character of the *reality* under investigation are three fundamental components of the structure of *Capital*. The first two are subordinated to and implied by the third. The external organization and the literary treatment of the subject-matter adequately express the *character* of the *reality* that has been investigated, i.e. comprehended and scientifically explained. Consequently, the structure of *Capital* does not and *could not* follow any *single* scheme. If the universal scheme of *Capital's* construction were the progression from essence to appearance, from the hidden concealed kernel to the phenomenal appearance,[22] then the overall organization of the work, which does follow this scheme, would radically differ from the exposition of details which (frequently) proceeds in the very opposite direction, from the phenomenon to the essence. Marx analyses a commodity, the simplest social form of labor product under capitalism, first in its *phenomenal* form, i.e. as exchange value, and only then does he proceed to examine its essence – i.e., value.

Marx introduces his work by an analysis of a *commodity*. What is a commodity? A commodity is an external object and at first glance a simple thing. It is the 'magnitude' with which the man of the capitalist society has his most frequent daily contact. It is the self-evidence of this world. But in the course of his analysis, Marx proves that a commodity is banal and trivial *only* at first glance, whereas in reality it is mystical and mysterious. It is not only a sensory–intuitive object but a sensory–transsensory thing at the same time.

How does Marx know that a commodity is 'the concrete form of labor product', 'the simplest economic concretum', and 'a form of a cell' which in a *concealed, undeveloped, abstract* way contains all the basic determinations of capitalist economy? The finding that a commodity is the elementary economic form of capitalism can become the *starting* point of a scientific explication only if the entire subsequent process of presentation *substantiates* the appropriateness and necessity of this starting point. In order to *start off* with a commodity as a totality of capitalism's abstract and undeveloped determinations, Marx already had to *know* capitalism's developed determinations. A commodity could become the starting point of a scientific presentation only because capitalism was known in its entirety. From the methodological standpoint, this amounts to exposing the dialectical connections between the element and totality, between the undeveloped germ and the fully-fledged functioning system. The appropriateness and the necessity of a commodity as the *starting* point for analysing capitalism is substantiated in the first three books of *Capital*, i.e. in its *theoretical* part. The second question is: Why did Marx arrive at this knowledge precisely in the second half of the 19th century? This is answered in *Capital's* fourth book, *Theories of Surplus Value*, i.e. in its literary–historical part where Marx analysed the decisive periods in the development of modern economic *thought*.

From the elementary form of capitalist wealth and from an analysis of its elements (the two-fold character of labor as a unity of use-value and value; exchange-value as the phenomenal form of value; the two-fold character of a commodity as an expression of the two-fold character of labor), the investigation proceeds to the *real* movement of commodities (to commodity exchange). It depicts capitalism as a system *formed* by the movement of the 'automatic subject' (value), with the system as a whole appearing as a system of exploiting another's work, as one that reproduces itself on a larger scale, i.e. as a mechanism of dead labor ruling over live labor, object ruling over man, product over its producer, the mystified subject over the real subject, the object ruling over the subject. Capitalism is a dynamic system of total reification and alienation, cyclically expanding and reproducing itself through catastrophes in which 'people' act behind masks of officers and agents of this mechanism, i.e. as its own components and elements.

A commodity which at first appeared as an external object and a trivial thing plays the role of a mystified and mystifying *subject* of capitalist economics whose *real* movement forms the capitalist system. Whether the real subject of this social movement is value or commodity,[23] the fact is that three theoretical volumes of Marx's work trace the 'odyssey' of this subject, i.e. *they describe the structure of the capitalist world (its economy) as formed by the subject's real movement.* To investigate the real world of this subject means: (1) to determine *the laws* of its movement, (2) to analyse the *real individual shapes or formations (Gestalten)* that the subject forms in and for its movement, and (3) to present a picture of this movement as a whole.

Only now have we developed the prerequisites for scientifically comparing and critically analysing Marx's *Capital* and Hegel's *Phenomenology of the Spirit*. Both Marx and Hegel anchor the construction of their respective works in a *common* metaphorical motif current in the cultural milieu of their time. This contemporary motif of literary, philosophical and scientific creation is that of an *'odyssey'*. To know himself, the subject (be it the individual, individual consciousness, spirit, collectivity, etc.) must *journey* through the world and get to know the world. Cognition of the subject is possible only on the basis of this subject's own activity in the world. The subject gets to know the world only by actively interfering in it, and only through actively transforming the world does he get to know himself. Cognition of *who* the subject is means cognition of the subject's activity in the world. But the subject who returns to himself after having journeyed through the world is different from the subject who had started out on the journey. The world which the subject has traversed is a different, changed world, because even the subject's journey has left its mark and traces in it. But in addition, the world *appears* different to the subject as he returns, because accumulated experience has influenced his way of *seeing* the world and has modified his attitudes to it in a certain way, in degrees ranging from conquering the world to resigning in it.

Rousseau's 'history of the human heart' (*Emile or Education*), the German *Bildungsroman* in its classical form of Goethe's *Wilhelm Meister* or in the romantic form of Novalis' *Heinrich von Ofterdingen*, Hegel's *Phenomenology of the Spirit* and Marx's *Capital* all employ the 'odyssey' motif in different realms of cultural creation.[24]

The odyssey of the *spirit* or the science of the *experience of consciousness* is not the only or universal type, but just one of the ways of 'realizing' an odyssey. Whereas the *Phenomenology of the Spirit* is 'the path of natural consciousness which presses forward to true knowledge' or 'the path of the soul that traverses a series of its own forms of embodiment as so many stations' so that 'through the complete experience of its own self' it may arrive 'at the cognition of what it is in itself',[25] *Capital* turns out to be the 'odyssey' of concrete historical praxis which proceeds from the elementary labor *product* through a series of real formations in which the practical—spiritual activity of people in production is objectified and fixed, to conclude its journey not in the cognition of what it is in itself, but rather in a *revolutionary* practical action based on this cognition. For the odyssey of the *spirit*, real forms of life are only indispensable moments in the evolution of a *consciousness* progressing from ordinary consciousness to absolute knowledge, from consciousness of the everyday life to the absolute knowledge of philosophy. In absolute knowledge, the movement is not only completed, but also closed. Cognition of oneself is activity, but of a particular kind: it is an intellectual activity, i.e. philosophy, performed by the Philosopher (that is, as a contemporary French commentator aptly puts it, by *le Sage*).

The opening paragraph of *Capital* emphasizes precisely the *materialist* character of philosophy, the basis for scientifically investigating economic problems: precisely because it is *not* an odyssey of the spirit, it does not start with consciousness. Rather, it is an odyssey of a concrete historical form of praxis, and *therefore* it starts with a commodity. A commodity is not only a trivial and a mystical thing, a simple thing with a two-fold character, an external object and a thing perceptible by the senses. Also, and above all, it is a sensory-*practical* thing, a *creation* and an *expression* of a particular historical form of *social labor*. We can now formulate the original question about the internal relation of *Capital's* beginning and its end, of commodities and classes, as follows: What is the connection between commodities *as an historical form of people's social labor*, and the practical—spiritual activity of social groups in production, i.e. of classes? Marx starts out with the historical form of the social *product*, describes the laws of its movement, but his entire analysis *culminates* in finding that *these* laws express in a certain way the social relations and the production *activity* of *producers.* To depict the capitalist mode of production in its totality and concreteness means to describe it not only as a lawlike process in itself, i.e. as a process carried out without, and independently of, human consciousness, but also as a process whose laws deal with the way people are *conscious* of both the process itself and of their position in it.[26] Marx's *Capital* is not a theory but a theoretical *critique* or a *critical* theory of capital. Besides describing objective formations of capital's social movement and the *forms of consciousness* of its agents that correspond to these formations, and besides tracing the objective laws of the system's *functioning* (complete with its disturbances and crises), it also investigates the genesis and the process of forming the *subject* who will carry out a *revolutionary* destruction of the system. A system has been described in its totality and concreteness if the immanent laws of its movement and destruction have been described. *Recognizing, and becoming conscious of* the character of the system as one of exploitation is an indispensible condition for the odyssey of one historical form of praxis to culminate in a *revolutionary* praxis.[27] Marx has described this recognition as an epoch-making consciousness.[28]

MAN AND THING, OR THE CHARACTER OF ECONOMICS

Our critical analysis has demonstrated both that individual reified aspects of economics are real moments of reality, and that these reified moments are fixed in theories and ideologies where they appear in different stages of intellectual development as *'care'*, *'homo oeconomicus'*, or as the *'economic factor'*. These guises of economics are both subjective and objective, they both exist for consciousness and reveal economics in particular ways. We have been searching them for approaches that would allow to detect the proper character of economics. Apart from being a critique of *concepts* and of real *reified* economic formations, our analysis has also uncovered certain aspects of the character of economics itself. The following analysis will retrospectively shed more light on individual reified moments of reality.

Social Being and Economic Categories

If economic categories are 'forms of being' or 'existential determinants' of the social subject, then their analysis and dialectical systematization uncovers social *being.* It is spiritually reproduced in the dialectical unfolding of economic categories. This shows once again why economic categories in *Capital* cannot be systematized in the progression of factual historicity or of formal logic, and why dialectical unfolding is the only possible logical construction of social being.

It is incorrect to state that every economic category of Marx's *Capital* is a philosophical category as well (H. Marcuse). It is true, though, that only a philosophical analysis which extends beyond the framework of a specialized science and which discovers *what reality is* and *how socio–human reality is formed* will enable one to comprehend the principle of economic categories and thus provide the key for their critical analysis. Economic categories do not tell what they are, but affect the life of society more like mysterious hieroglyphs. The statement that social being is formed by interest, wages, money, rent, capital and surplus value will consequently sound arbitrary and absurd, and rightly so. As long as it was tracing the *movement* of economic categories, economic science never questioned what these categories *are*, and never even considered looking for the internal connections between these categories and social being. On the other hand, a *conception of reality* entirely different from that of classical economics was needed in order to discover this connection. The analysis of a certain reality, in this case of the economics of capitalism, is the work of science, of political economy. However, to be really scientific and not to hover over the fringes of science (as does Moses Hess' philosophizing about economic phenomena, or the doctrinaire systematization of ideas about economic reality, found in vulgar economics), it has to be anchored in a true conception of social reality, one that neither is nor can be a matter for any specialized scientific discipline.

Economic categories are not philosophical categories, yet the discovery of what they are, and thus also their critical analysis, necessarily starts from a philosophical conception of reality, science, and method. The critical analysis which demonstrates that economic categories are not what they appear to be and what uncritical consciousness presents them as, and which exposes their concealed inner kernel, also has to prove – if it wants to maintain a scientific level – that their categorical appearance is a *necessary* manifestation of their concealed essence. This process, in which the pseudo–concrete is abolished in order to demonstrate it as a *necessary* phenomenal form, transcends in no way the framework of philosophy (i.e. of Hegel). Only the proof that economic categories are historical forms of man's objectification and that as products of historical praxis they can be transcended only by practical activity, will indicate the limits of philosophy and the point where revolutionary activity takes over. (The reason why Marx followed in the footsteps of classical science and rejected romanticism, though at first glance it should have been the other way around, is this: while the classics presented an analysis of the objectual world, romanticism was only a protest against this world's inhumanity, and in this sense was also its product, i.e. something derivative and secondary.) The analysis of economic categories is not presuppositionless: it assumes a conception of reality as a practical process of producing and reproducing the social man. Such an analysis discovers in economic categories basic or elementary forms of objectification, i.e. of the objective existence of man as a social being. It is of course true, and regardless of all romantic protestations classical economics was correct on this point, that economics as a system and as a totality requires and forms a man that suits its own perspective. It incorporates man into its system to the extent to which he posesses particular characteristics, to the extent, that is, to which he is reduced to the 'economic man'. But economics is the objectified and realized unity of subject and object, it is the elementary form of objectification, of man's objectified practical activity, and precisely therefore this relationship creates not only objective social wealth but also subjective qualities and capacities of people. 'Not only do the objective conditions change in the act of reproduction, e.g. the village becomes a town, the wilderness a cleared field, etc., but the producers change, too, in that they bring out new qualities in themselves, develop themselves in production, transform themselves, develop new powers and ideas, new modes of intercourse, new needs and new language.'[29]

Economic conditions express 'forms of being' or 'existential determinations' of the social subject only in their totality which, far from being just a pile of categories, forms a dialectical construction determined and constituted by an 'all-controlling power', i.e. by that which forms the universal 'ether of being', as Marx puts it. All other categories taken by themselves and in isolation express only its *individual* facets and partial aspects. Only when these categories dialectically unfold and when their construction suggests the internal organization of a given society's economic structure, only then does *each* of the economic categories acquire its own real sense, only then does it turn into a concrete historical category. It is then possible to discover in each of these categories the following — either in essence (for basic economic categories), or in a certain aspect (for auxiliary categories):

(1) a certain form of the socio-historical objectification of man, since as Marx remarks, production is in essence the objectification of man;[30]

(2) a certain concrete historical level of the subject—object relationship; and

(3) the dialectics of the historical and the trans-historical, i.e. the unity of ontological and existential determinations.

If this new concept of reality (the discovery of praxis and of revolutionary praxis) provides the grounds for exposing the character of economic categories and for analysing them, then social reality can be in turn constructed from these categories. The economic structure of society is spiritually reproduced in the system of economic categories. It is then also possible to discover what economics actually is, and to distinguish that which amounts to reified and mystified forms of economics or to its necessary external phenomena from economics in the proper sense of the word. Economics is not only the production of material goods; it is the totality of the process of producing and reproducing man as a socio—historical being. Economics is the production of material goods but also of social relations, of the context of this production.[31]

What bourgeois and reformist critics take to be the 'speculative', 'messianic', or 'Hegelizing' part of *Capital* is only an external expression of the fact that beneath the world of objects, beneath the movement of prices, commodities and of different forms of capital, whose *laws he expressed in exact formulas*, Marx exposed the objective world of social relations, i.e. the object—subject dialectics. *Economics is the objective world of people and of their social products; it is not the objectual world of the social movement of things*. The social movement of things which masks social relations of people and of their products is one particular, historically transient form of economics. As long as this historical form of economics exists, i.e. as long as the social form of labor creates exchange-value, there also exists a real prosaic mystification. When mystified, particular relations into which individuals enter in the course of producing their social life appear inverted, as social relations of things.[32]

In all these manifestations, both economics as a whole and individual economic categories show themselves as a *particular* dialectics of persons and things. Economic categories, which in one respect fix social relations of things, incorporate within themselves people as *agents* of economic relations. The analysis of economic relations is a *twofold* critique: First, it demonstrates the failure of earlier classical economic analyses to adequately express *this* movement; in this sense the critical analysis is a *continuation* of classical economics: it rectifies the latter's contradictions and shortcomings and presents a more profound and a more universal analysis. Second, and this is where Marx's theory is a *critique* of economics in the proper sense of the word, it exhibits the *real movement of economic categories as a reified form of the social movement of people*. *This* critique *discovered that categories of the social movement of things are necessary and historically transient existential forms of the social movement of people*. Marxist economics thus originates as a twofold critique of economic categories, or, stated positively, as an analysis of the *historical dialectics of people and things in production*, grasped as *the socio—historical production of objective wealth and of objective social relations*.

In capitalist economies, things and persons become interchangeable. Things are personified and persons are reified. Things are invested with a will and a consciousness, i.e. their movement is conscious and willful, and people turn into agents and executors of the movement of things. The will and consciousness of people are determined by the objective course of things: the movement of things employs the will and consciousness of people as *its own* medium.

The lawlike character of things that follows from their social movement is transposed in human consciousness as an aim and an objective; the subjective purpose is objectified and functions independently of individual consciousness as a tendency or a raison d'être of the thing. The 'Raison d'être, inner drive, tendency' of value and commodity production appears in the consciousness of the capitalist, which had mediated this raison d'être, as his own conscious intention and purpose.[33]

If one traces and formulates the lawlike character of the social movement of things, for which man (homo oeconomicus) is merely an agent or a character mask, it becomes immediately obvious that this reality is only a real semblance. At first sight, man in the economic production might appear as a mere personification of the social movement of things and as a conscious executor (agent) of this movement.[34]

Further analysis, however, *abolishes this real semblance and proves that the social movement of things is an historical form of contact among people, and that reified consciousness is only one historical form of human consciousness.*

Economic categories from which social being has been constructed and which amount to the existential forms of the social subject are not therefore expressions of the movement of things, or of social relations among people, that would be severed from the people themselves and from their consciousness. Rather, fixed in economic categories are social relations of production which traverse human consciousness but are independent of it, i.e. *make use* of the consciousness of individuals for *their own* existence and for *their own* movement. The capitalist is a social relation decked out with a will and a consciousness, mediated by things, and manifested in *their* movement.[35]

Social being *determines* the consciousness of people but that does not imply that it is adequately uncovered in their consciousness. In their everyday utilitarian praxis, people are more prone to become aware of social being in its separate aspects and in its fetishised forms. How is man's social being exposed in economic categories? Does one disclose social being by translating it into the corresponding economic category, such as capital, land tenure, small-scale production, monopoly, etc., or into the facticity of conditions and data of economic history? In such a translation, certain forms or isolated moments would be substituted for social being, so that the assignment of cultural formations to being, conceived in this way, could not go beyond vulgarization, a thousand times though it might assert that the relationship between 'economics' and 'culture' is of course understood in a 'mediated' and 'dialectical' fashion. The approach is vulgar not for a lack of mediation, but in its very manner of grasping social being. Social being is no substance, rigid or dynamic, and neither is it a transcendental entity existing independently of subjective praxis. Rather, it is *the process of producing and reproducing social reality, i.e. the historical praxis of mankind and forms of its objectification.* Economics and economic categories are on the one hand incomprehensible without objective praxis and without answering the question, how is social reality formed; but on the other hand, inasmuch as they are the basic and elementary forms of man's objectification, they are the constitutive elements of social being. 'When we consider bourgeois society in the long run and as a whole', and this is how Marx sums up the connections among social being, praxis, and economics, 'then the final result of the process of social production always appears as the society itself, i.e. the human being itself in its social relations. Everything that has a fixed form, such as the product, etc., appears as merely a moment, a vanishing moment, in this movement. The direct production process itself here appears only as a moment. The conditions and objectifications of the process are themselves equally moments of it, and *its only subjects are* the individuals, but *individuals in mutual relationships* which they equally reproduce and produce anew. The constant process of their own movement, in which they renew themselves even as they renew the world of wealth they create'.[36] Social being is not 'contained' in economic categories and in their dialectical organization, it is *fixed* there. A theoretical analysis will therefore expose social being in the system of economic categories only after 'dissolving' their fixed attachments, once it conceives of them as an expression of people's *objective praxis* and of their interconnected *social* relations at a particular *historical* stage of development.

Philosophy of Labor

So deeply is the connection between economics and labor rooted in ideas of science and of everyday consciousness that nothing seems easier than to analyse labor in order to grasp the character of economics, or conversely, to slash through the thicket of economics in order to comprehend labor. This apparent self-evidence is, however, misleading. Far from guiding the investigation toward an analysis of labor, it *smuggles in* a different problem and orients science toward describing and analysing work processes or toward historical and systematic surveys of work activities, generalized under a 'definition of work'. These definitions describe or generalize *work activity*, or work in its empirical form, but leave problems of labor untouched. Sociology of work, psychology of work, theology of work, physiology of work or economic analyses of work, and the corresponding sociological, psychological, economic, etc. concepts all deal with and fix *particular aspects* of labor. However, they take the *central* question, of what labor is, for granted as an unexamined and uncritically accepted assumption (and so-called scientific investigation is consequently based on an unscientific bias), or they consciously excise it out of science as a 'metaphysical question'.[37] Sociological definitions of work that attempt to avoid abstractness and to exclude metaphysics present a generalized description of *work processes* or of *work activity* but never penetrate to problems of labor. From the standpoint of sociology of work it is *a priori* impossible to get at the problem of labor. Though it may seem that nothing is more familiar and trivial than labor, this familiarity and triviality turns out to be based on a *substitution*: the everyday image of work and its sociological systemization does not deal with the essence and with the universal character of labor; rather, the term 'work' connotes work processes, work routines, different kinds of work, etc. A 'philosophy of labor' therefore does not reflect upon sociological definitions and findings or data gathered by anthropologists, psychologists and physiologists. Its task is not to *generalize* partial findings of various sciences, let alone to present an apology for a particular historical form of labor.[38] Philosophy does not offer an analysis of work processes in their totality or in their historical development, but rather deals with a single question: What is labor?

Except, does the term 'philosophy of labor' not misuse the notion or the concept of philosophy? Why does an analysis of labor require a philosophical examination, why can it not be performed within the framework of a specialized science? Or is perhaps this expression parallel to those of 'philosophy of games', 'philosophy of language' or 'philosophy of art', and does it then denote yet *another* discipline of the humanities investigated from the philosophical standpoint?

The problematique which we are subsuming under the expression 'philosophy of labor' has appeared in important historical junctures of modern European thought: in the Renaissance (G. Manetti, Pico della Mirandola, Carolus Bovillus), in Hegel's philosophy, and in Marx. Problems of a 'philosophy of labor' are an early aspect of the question: 'Who is man?' To avoid possible misunderstanding, we have to add this: The problem of labor as a *philosophical* question accompanies questions concerning the being of man only *providing* that the question 'Who is man?' is conceived as an ontological one. The 'ontology of man' *is not* an anthropology.[39] The problem of labor as a *philosophical* question and as a *philosophy of labor* is based on an ontology of man. The connection of labor with the philosophical problematique of above-mentioned currents of thought is thus more than just factual. The incredulous observation that no philosophy of labor has been developed since Marx's time[40] is meaningful only when coupled with another observation, that materialist philosophy is also the 'latest' 'ontology of man', in that it has not been rendered obsolete by history.[41]

In its essence and universality, labor is not work activity or a job that man does, and which would in turn influence his psyché, his habits and thinking, i.e. limited aspects of human being. Labor is a *happening* which permeates man's *entire* being and constitutes his specificity. Only such thinking as that which discovered that something essential is happening to man and to his being in the process of working,[42] and which beheld the *necessary internal* connection between the questions, 'What is labor?' and 'Who is man?', could initiate a scientific investigation of *all* forms and manifestations of labor (including its economic problems), as well as of all forms and manifestations of human reality. Inasmuch as labor is a doing and a happening in which *something* happens with man and with his being, as well as with his world, the interest of philosophy understandably concentrates on clarifying the character of this 'happening' and of this 'doing', on discovering the secret of this 'something'. This problematique is frequently dispatched with the suggestion that labor is the point where causality and teleology conjoin and unify in a specific manner, or where the animal is transformed into the human, i.e. that it is the locus of man's *genesis*.[43] Correct though such analyses may be, they amount to no more than *partial* knowledge. They bypass the problematique which is revealed in finding that in addition to the dialectical pairs which this analysis does list and

investigate — causality–teleology and animality–humanity — *other* dialectical pairs can be discovered in the happening of labor too, such as necessity
and freedom, the particular and the general, the real and the ideal, the
internal and the external, theory and praxis, man and nature, etc.[44] Does
perhaps the pair causality–teleology occupy a privileged position in the
investigation of the problem of labor, or did the investigative procedure
omit other dialectical pairs because it was not systematic enough? How can
the *completeness* of a systematic series of dialectical pairs be assured? And
would it follow that labor is a privileged category on the basis of which an
entire system of dialectical categories can be constructed, or that a system
of dialectical categories is to be anchored in the concept of labor, as its
natural and necessary center?

This analysis is not usually criticized for being unsystematic, when it
focuses on and thus privileges one or two dialectical pairs from a whole
range of them. The one-sidedness of this approach does, however, have one
fundamental shortcoming: the arbitrary and one-sided selection amounts to
an incapacity to scientifically *formulate* the problem, and makes it impossible to penetrate to the *essence* of the question. Inasmuch as the concept of
labor is exhausted or characterized by one or two dialectical pairs, or by
some incomplete set of them, the elements of these pairs will stand out as
categories, and the analysis of labor will turn either into an analysis or
systematization of categories, complete or incomplete, or into a particular
example or instance (causality and teleology, etc.) used to clarify these
categories. The critique of the shortcomings of *partial* analyses thus does
not call for their completion, for generating a systematic series of partial
analyses, but rather highlights the question: *Wherein lies the specificity of
those dialectical pairs* which are to describe labor?

The general characterization of labor as a happening and a *doing* in which
something happens with man and his being has to have some connection
with the dialectical pairs employed to describe labor. There is no specific
connection between the causality–teleology pair on the one hand, and
other pairs, such as the particular and the universal, freedom and necessity
or real and ideal, on the other hand, except for their general dialectical
character. If there were a connection between the *dialectics* of *these* pairs
and the *happening* of labor, would this connection not uncover a dialectics
of the happening and the happening of dialectics? That is, will the *character*
of the happening of labor and the *content* of the dialectics not be specified
in the pairs used to describe labor? Dialectical pairs can describe labor and
its happening *adequately*, as long as this happening is revealed in their

dialectics as indeed *dialectics.* And should the analysis of *dialectical* happening of labor be internally connected with the *being* of man, then the happening of labor will simultaneously expose man's *specificity.*

The specificity of man's being is frequently illuminated by contrasting it with the being of a beast or with the being of things. What makes man different from a stone, a beast or a machine? As a dialectician, Hegel pinpointed the difference between man and beast in the area which they both essentially share: in the sphere of animality. Harnessing the animal craving[45] and interposing a mediating element – labor – between this craving and its satiation is not only a process of transforming animal craving into human craving,[46] a process through which man is born, but it is also an *elementary model of dialectics itself.* The transformation of animal craving into human craving, the humanizing of craving on the basis and in the process of labor, is only *one* aspect of the happening of labor. In other words: the access to the happening of labor which we gained by distinguishing animal and human cravings will lead to *grasping* this happening, providing however that the happening will not be viewed as a unique or an isolated metamorphosis and that it will be exposed as metamorphosis in general. Labor is a happening, and what happens is a metamorphosis, i.e. dialectical mediation. The dialectical mediation of this happening does not *balance* opposites, nor are opposites constituted in an *antinomy.* Rather, in the process of *transformation* a unity of opposites is *formed.* A dialectical metamorphosis is a metamorphosis in which *new* is *formed.* A dialectical metamorphosis is the genesis of what is qualitatively new. The same act of mediation in which animality begets humanity and in which animal craving is transformed into humanized craving, into the craving for being craved, i.e. the craving for recognition, also forms the *three-dimensionality of human time*: only a being which transcends the nihilism of its animal craving in labor will in the act of harnessing its craving uncover a *future* as a dimension of its being. Through work, man *controls* time (whereas the beast is exclusively *controlled* by time), because the being that can resist immediate satiation of its craving and can 'actively' harness it forms a present as a function of the future, while making use of the past. *In its doing it uncovers the three-dimensionality of time as a dimension of its own being.*[47]

Having transcended the level of instinctive activity, and having turned into an exclusively human doing, labor transforms the given, the natural and the non-human, and adapts it to human needs even as it realizes human intentions in material of nature. Nature thus appears to man in a double light: it stands out as a power and an objectivity that has to be respected, whose laws have to be recognized so that man may use them to his own advantage, yet it sinks to the level of mere material in which human intentions are realized. Man gives full rein to natural forces that exist *independently* of him to act to his own advantage and in his own interest, but he also objectifies himself in nature and in the material of nature, thereby degrading it to mere material for *his own* meanings. (We shall deal in greater detail with this problem of mutual activity and passivity between man and nature in Chapter Four.) Labor is both a transformation of nature, and a realization of human meanings in it. Labor is a happening or a doing in which the unity of man and nature is constituted in a certain way, on the basis of their mutual transformation: man objectifies himself in labor and the object is torn out of its original context, adapted and processed. Through labor, man is objectified and the object is humanized. In humanizing nature and in objectifying (realizing) meanings man forms a human *world*. Man lives in a world (of his own artifacts and meanings), whereas the animal is tied to conditions of nature.

The constitutive element of labor is *objectivity*. Objectivity of labor means, first, that the result of labor is a product which has *duration*, that labor has sense only if it 'constantly undergoes a transformation: from being motion (*Unruhe*), it becomes an object without motion; from being the labor working, it becomes the thing produced (*Gegenständlichkeit*)'[48] — that is, if it appears as a cycle of activity and duration, of movement and objectivity. When the labor process is over, the product of labor in the broad sense of the word endures as its end result and its incarnation. What had appeared as progression in *time* during the labor process, appears as the condensation or *abolition* of the time succession, as inertness and duration, in the *product* of labor. In the labor process, results of *past* labor are transformed while realizing intentions of the *future*. The three-dimensionality of human time as a constitutive dimension of man's being is anchored in labor as man's *objective doing*. The three-dimensionality of time and the temporality of man are based on objectification. *Without objectification there is no temporality*.[49] As objective doing, labor is a special mode of identity of time (temporality) and space (extension), as two fundamental dimensions of human being, of a specific form of man's movement in the world.

Second, the objective character of labor is a manifestation of man as a *practical* being, i.e. of an *objective* subject. In labor, man leaves behind something permanent, something that exists *independently* of individual consciousness. The existence of objectified artifacts is a prerequisite of history, i.e. of continuity in human existence. In this context it becomes clear why a profound and realistic view of socio-human reality appreciates the *tool* more than the intention, and emphasizes its *central* position in stating that the tool is 'reasonable mediation' between man and the object. In intellectual history, this line has been advocated by those philosophers who emphasized the significance of the human hand and its connection with man's reason. Anaxagoras has said that 'it is the possession of hands that makes man the wisest of living things'. Aristotle, and G. Bruno after him, have described the hand as 'the tool of tools'. Hegel completes this line. By contrast, romantic philosophy expresses its disdain for technology and utopically denounces a world in which 'man is lost in his tools'.

There is a widespread opinion that man is the only being aware of its mortality: only he faces a future opening up ahead, with death at its end. The existentialist interpretation of this opinion idealistically distorts it. From the finitude of man's existence it infers that objectification is a form of *flight* from authenticity, namely from being-toward-death. But man knows his mortality only because he organizes time, on the basis of labor as objective doing and as the process of forming socio—human reality. Without this objective doing in which man organizes time into a future, a present and a past, man could not *know* his totality.

Labor and Economics

We had expected our analysis of labor to clarify the character of economics but it has led us to the 'ontology of man' instead. This digression was a necessary detour which brought us closer to the problem. The philosophical analysis might not have told us what economics is, but it did uncover certain fundamental features of man's being. On the other hand, it has turned out that in order to grasp labor as labor, as distinct from work activity, work routines and from particular historical forms of labor, it has to be interpreted as a specific happening or as a specific reality that constitutes and permeates man's entire being. Earlier analytical attempts described labor by using dialectical pairs such as causality—teleology, animality—humanity, subject—object, etc., with labor itself standing out as an 'active center' in which the dialectical unity of these pairs were *realized*. They outlined the essential features of labor but did not capture its specificity. Earlier characteristics included man's doing in general but did not distinguish among its different kinds.

The medieval ruler would never have considered ruling as labor, nor would he have thought of himself as working when involved in political decision-making. And as Marx noted, Caesar or Aristotle would have been positively insulted by the very title 'laborer'. Does this mean that political activity, science and art *are not* labor? A sweeping negative answer would be just as incorrect as the assertion that science, politics and art indeed are labor.[50] Where is the limit of labor, or the measure of its distinctiveness? Or do perhaps the mentioned kinds of man's doing amount to labor only under some circumstances and not under others?

Art has *always* been considered a human activity and a human doing *par excellence*, a free creation distinct from labor. Hegel posits genuine labor in the place of artistic creation which had been the only kind of praxis for Schelling. Hegel's is both a more democratic and a more profound view of human reality. This distinction should not, however, obscure the *other* side of the problem. For Schelling, as for Augustin Smetana and Edward Dembowski, artistic creation was a *free* 'praxis', i.e. a kind of human doing that is not subject to outside necessity, and is explicitly characterized by 'independence on extraneous purposes'. Human doing is thus divided into two areas: in one it is performed under pressure of necessity and is called labor, in the other it is realized as free creation and is called art.[51] This distinction is correct insofar as it succeeds in capturing the specificity of labor as such an objective doing of man which is instigated and constitutively determined by *extraneous* purpose, whose satisfaction is dubbed natural necessity or social obligation. Labor is human doing involved in the realm of *necessity*. Man labors insofar as his doing is provoked and determined by the pressures of outside necessity, the satisfaction of which supports his existence. One and the same activity can be both labor and not-labor, depending on whether or not it is performed as a natural necessity, i.e. as a necessary prerequisite of existence. Aristotle did not labor. A professor of philosophy does, because his translations and interpretations of Aristotle's *Metaphysics* are an *occupation*, i.e. a socially conditioned necessity to acquire material means of livelihood and existence.

Dividing human doing into labor (the realm of necessity) and art (the realm of freedom) captures the problem of labor and not-labor only *approximately* and in some of its aspects. This distinction is based on a definite *historical* form of labor as an unexamined and thus uncritically accepted assumption, which leads to *petrifying* a particular historically *generated* division of labor into that which is material—physical, and that which is spiritual. This distinction conceals another essential feature of the specificity of labor, namely that labor is a human doing which *transcends* the realm of necessity *and forms within it* real prerequisites of human *freedom*,[52] even without leaving it.

Freedom does not disclose itself to man as an autonomous realm, *independent* of labor and existing beyond the boundaries of necessity. Rather, it grows out of labor which is its necessary prerequisite. Human doing is not split into two autonomous realms, mutually independent and indifferent, one of which would incarnate freedom and the other constitute the arena of necessity A philosophy of labor, i.e. of an objective human doing *through which*, in the happening of necessity, real prerequisites of freedom are *formed*, is consequently also a philosophy of not-labor. The objective doing of man that transforms nature and imprints into it his meanings is a *unified* process which, though performed out of necessity and under pressure of extraneous purposiveness, also realizes the prerequisites of freedom and free creation. The splitting of this unified process into two *seemingly* independent realms does not follow from the 'nature of the matter' but is historically a transient state. As long as consciousness is a captive of this split, it will not behold its *historical* character and will *juxtapose* labor and freedom, objective activity and imagination, technology and poetry as two independent ways of satiating the human drive.[53]

On the other hand it is natural that the romantic absolutization of dreams, imagination and poetry will accompany, as its faithful *alter ego*, any 'fanaticism of labor' – i.e. any historical form of production in which the unity of necessity and freedom is realized through separating labor from joy (pleasure, bliss, happiness), or as a unity of opposites which are *personified* in *antagonistic* social groups.[54] Such human doing which is determined only by internal purposiveness and does not depend on natural necessity or social obligation is not labor but free creation, *irrespective* of the realm in which it is realized. The real realm of freedom thus begins beyond the *boundaries* of labor, although precisely labor forms its indispensible historical basis. 'The realm of freedom actually begins only where labor which is determined by necessity and mundane considerations ceases; thus in the very nature of things it lies beyond the sphere of actual material production.'[55]

These considerations dispel the impression that labor belongs in and of itself to economics, or that it is characteristically a 'natural' economic concept. So far we have found nothing economic about labor. We have, however, reached the point of exposing both the *internal* connection between economics and labor, and the character of economics. Economics is neither exclusively the realm of necessity nor the realm of freedom; it is a sphere of human reality in which a unity of necessity and freedom and of animality and humanity is *historically* formed. Economics is the realm of necessity (of the objective doing of labor) in which historical prerequisites of human freedom are *formed*. Our analysis of labor has led us to two important findings about economics. The first concerns the *origin* of economics. Because we approached the investigation of economics from an analysis of labor, economics turned out to be primarily not a *ready-made* economic structure of reality, an *already-formed* historical base and unity of production forces and production relations, but rather a socio—human reality *in the process of formation*, a reality based on man's *objective—practical* doing. Second, we established the *position* of economics in the socio—human reality: economics occupies the *central* position in this reality, because it is the arena for the historical metamorphosis through which man is formed as a reasonable being and a social creature, through which man is humanized. Economics is located at the point where animality is humanized and where the unity of necessity and freedom is realized. In this sense, economics appears as the *conjunction* of human relations and the source of human reality.

Two extreme opinions will serve to illustrate misunderstandings concerning the position of economics in the system of human reality. Schelling, who generally sought a 'higher necessity' and a 'true reality' behind empirical phenomena, was so shocked by the supremacy of 'economic interest' in his time, that he could not extricate himself from the bondage of these reified empirical facts and in this instance did not even search for the 'true reality'. What is economics, Schelling asked: commerce, sugar beet, breweries and cattle raising?[56] The second extreme is the opinion which places economics on the *periphery* of human reality and takes it to be a sphere concerned exclusively with physical needs, a sphere of satiating the elementary needs of man as a physiological, biological, animal being. Economics is consequently seen as playing a decisive role only in *extreme* situations when all human interests are cast aside and when all that is left is the urgent need to eat, be warm, be clad. Economics would become a determining factor at times of famine, war, natural catastrophes. When does man live by economics, when is he determined by economics? When he has nothing to eat and is cold, our author says.[57]

If we inquire about the relationship between labor and the forming of socio—human reality, we shall discover nothing economic about labor. Labor as man's objective doing in which the socio—human reality is *formed* is labor in the *philosophical* sense. On the other hand, labor in its economic sense is the creator of a specific historical *form* of wealth. From the economic standpoint, labor turns out to be the regulator and the active structure of social relations in production. As an economic category, labor is a socio—productive activity that forms a specific kind of social wealth.[5][8]

Although labor in general is the presupposition for labor in its economic sense, the two are not identical. The labor that forms the wealth of the capitalist society is not labor in general but is rather a *particular* labor, it is the abstract—concrete labor, i.e. labor with a two-fold character, and only in *this* form does it belong in economics.

NOTES

[1] For example, Jean Domarchi writes that 'viewed from a historical perspective, the Marxist analysis is dialectical, and it portends what phenomenology would be'. *La revue internationale*, Paris 1945—6, pp. 154—67. Pierre Naville in the same issue answers Domarchi in his article 'Marx ou Husserl' and rejects the proposed symbiosis of Marxism and phenomenology. However, he falls victim to naturalist and mechanistic errors, and the discussion can therefore not be considered closed.

[2] P. Bigo, *Marxisme et humanisme: Introduction à l'oeuvre de Marx*, Paris 1954, p. 7.

[3] Ibid., p. 21.

[4] 'Marx's confrontation with philosophy resulted in the same conclusion as that with economists. Marxist political economy is above all an analysis of existence'. Ibid., p. 34. The general false interpretation leads this Thomist author in a number of places to hardly excusable errors and mystifications. Bigo describes Marx's critique of capitalist fetishism as a 'subjectivization of value'. In itself, this formulation might be considered just a little clumsy, providing it meant that Marxism translates the objective and reified character of social wealth into objective *activity*, i.e. points to the *genesis* of this reified result. Marxism could be associated with the attribute 'subjective' in this sense; i.e. as a theory exposing the historical *subject* of social wealth. Bigo, however, takes the 'subjectivization of value' to mean its de-objectification and spiritualization, as shown in his interpretation of Marx's critique of the Physiocrats. Marx did not criticize the Physiocrats' concept of value for its materialism, as Bigo believes, but for its naturalism, which of course is something entirely different. A more detailed critique of Thomist interpretations of Marx's work is presented in R. Garaudy, *Humanisme marxiste*, Paris 1957, pp. 61*ff.*, and L. Goldmann, *Recherches dialectiques*, Paris 1959, pp. 303 *ff.*

[5] Joseph Schumpeter has been a persistent proponent of this position. From his early essay *Epochen der Dogmen—und Methodengeschichte*, of 1914, up to his recent books, such as *Capitalism, Socialism and Democracy*, he has consistently divorced Marx the economist from Marx the philosopher. 'Wenn Marx in der Tat aus metaphysischen Spekulationen materielle Gedankenelemente oder auch nur die Methode erborgt hätte, so wäre er ein armer Schächer, nicht wert ernstgenommen zu werden. Aber er hat es nicht getan ... Kein metaphysischer Obersatz, nur — richtige oder falsche — Tatsachenbeobachten und Analyse hat ihn in seiner Werkstatt beschäftigt.' *Dogmengeschichte*, p. 81.

[6] 'A science of pure facts is absurd', Otto Morf correctly points out in *Das Verhältnis von Wirtschaftstheorie und Wirtschaftsgeschichte bei Karl Marx*, Basel 1951, p. 17. It follows from our previous exposition that Schumpeter's is but one of the possible interpretations of *Capital*, which Morf's critique misses.

[7] Most Marxist interpreters see it as a positive development, whereas Christian and existentialist Marxologists see it as a degeneration. Both instances stem from a false idea and a false interpretation of *Capital*.

[8] Marx's development is taken as a transition from a philosophical concept of alienation to the economic concept of commodity fetishism, or as a transition from the subject—object dialectics to the object—object 'dialectics'. (see 'Sur le jeune Marx', *Recherches internationales*, no. 19, Paris 1960, pp. 173f, 189.) These authors have not noticed that their 'transitions' result in an amazing transformation of Marx himself — into a positivist.

[9] In essence it amounts to the same case of idealism and utopia that Marx had exposed in the petit—bourgeois socialism of Proudhonists: 'Was diese Sozialisten von den bürgerlichen Apologeten unterscheidet, ist auf der einen Seite das Gefühl der Widersprüche des Systems, andererseits der Utopismus, den notwendigen Unterschied zwischen der realen und idealen Gestalt der bürgerlichen Gesellschaft nicht zu begreifen, und daher das überflüssige Geschäft zu übernehmen, den idealen Ausdruck, das verklärte und von der Wirklichkeit selbst als solches aus sich geworfene reflektierte Lichtbild, selbst wieder verwirklichen zu wollen'. Marx, *Grundrisse der Kritik der politischen Ökonomie*, Berlin 1953, p. 916; cf. also p. 160 (248 in English edition).

[10] 'When reason has been established as the rational organization of mankind, philosophy is left without an object'. 'The philosophical construction of reason is replaced by the creation of a rational society'. H. Marcuse, 'Philosophy and Critical Theory', in *Negations*, Boston 1968, pp. 135, 142. 'Critical theory' (Horkheimer, Marcuse) would abolish philosophy both ways: by realizing it as well as by transforming it into a social theory.

[11] Herbert Marcuse's *Reason and Revolution*, New York 1960, 2nd. ed., is based on this conception. The transition from Hegel to Marx is poignantly labelled 'From Philosophy to Social Theory' (pp. 251—57), and Marx's teaching is interpreted in a chapter called 'The Foundation of the Dialectical Theory of Society' (pp. 258—322). Marcuse had already formulated this conception back in the thirties, in his essays for Horkheimer's *Zeitschrift für Sozialforschung*. Judging from his later writings, the author became to a certain extent aware of the problematic character of his basic thesis, though he continued to maintain it: 'Marx's materialist 'subversion' of Hegel . . . was not a shift from one philosophical position to another, nor from philosophy to social theory, but rather a recognition that the established forms of life were reaching the stage of their historical negation'. H. Marcuse, *Reason and Revolution*, p. xiii.

[12] Especially Max Adler and, in a more vulgar form, Karl Kautsky. In all instances, Marxist sociology apparently has to be complemented by a non-Marxist philosophy, by Kant, Darwin or Mach.

[13] Esp. Karl Löwith, *From Hegel to Nietzsche*, New York 1964.

[14] See S. Kierkegaard, *The Present Age*, Oxford 1940.

[15] Marcuse, *Reason and Revolution*, p. 258.

[16] Ideas and terms such as the social question, the social novel, social poetry, etc., employed in the 19th century, are wholly foreign to materialist philosophy.

[17] This subjectivism finds its most radical expression in the opinion that there exists no social science, but only class consciousness. This opinion lends itself to a French pun: no 'science sociale', only 'conscience de classe'.

[18] Lenin, *Philosophical Notebooks*, in *Collected Works*, vol. 38, Moscow 1961, p. 180. It is known that Lenin had not read the *Phenomenology of the Spirit*. In the light of this simple fact, the argument of French philosophers over whether tracking down connections between *Capital* and *Logic* would be a manifestation of materialism, whereas tracking down connections between *Capital* and *Phenomenology* would be a manifestation of idealism, acquires a particularly ridiculous character.

[19] Jean Hyppolite, *Studies in Marx and Hegel*, New York 1969, p. 137. As we shall demonstrate later, the author never got beyond merely stating this connection. He mentioned some accidental points of contact between *Phenomenology of the Spirit* and *Capital*, which, however, are peripheral to the thing itself.

[20] The argument was provoked by Henryk Grossmann's paper, 'Die Änderung des ursprünglichen Aufbauplanes des Marxschen 'Kapital' und ihre Ursachen', *Archiv für Geschichte des Sozialismus und Arbeiterbewegung*, Leipzig, vol. 14 (1929): 305–338. However, Marx's manuscripts that were published later demonstrate that Grossmann proceeded from unwarranted premises; consequently, his dating of the suspected change in plan (Summer 1863) is incorrect, since Marx had a detailed plan of the *final* shape of *Capital* ready by the end of 1862. (See *Marx–Engels–Archiv* Moscow 1933, p. xii.) More recent authors, e.g. O. Morf in *Verhältnis von Wirtschaftstheorie*, accept Grossmann's theses with reservations or even fully (cf. e.g. Alex Barbon, 'La dialectique du Capital', *La revue internationale*, Paris 1946, no. 8, pp. 124ff), but none of them question the way the problem itself is postulated.

[21] Marx's letter to Engels, 31 July 1865 (*Werke*, vol. 31, p. 132). [*Selected Correspondence*, New York, 1942, p. 204.]

[22] Cf. Hyppolite, op. cit., p. 139.

[23] In *Capital*, Marx considers value to be the subject of this process, whereas in his polemic with Wagner, of 1879–80, he explicitly notes that commodity, not value, is the subject. See Marx, 'Randglossen zu Wagners Lehrbuch' ['Notes on Adolph Wagner' Marx: *Texts on Method*, tr. T. Carver, Oxford, 1975, pp. 179–219].

[24] As far as I know, the connection between Hegel's *Phenomenology of the Spirit* and the German *Bildungsroman* was first pointed out by Josiah Royce in his *Lectures on Modern Idealism*, New Haven 1919, pp. 147–49.

[25] G. W. F. Hegel, *Phenomenology of the Spirit*, New York and London, 1931, p. 135 (adapted).

[26] This is what Marx wrote about relations among people in exchange and production: 'At first, the relation practically exists. Then, however, since this is a matter of people, *the relation exists as a relation for them*. The way in which it exists for them, or in which their brain reflects it, follows from the very nature of these relations'. Marx, *Das Kapital*, Hamburg 1867, p. 38. This paragraph had been deleted from later editions.

[27] In his letter to Engels of 30 April 1868, Marx outlines the internal connections of the three volumes of *Capital*, and concludes: 'We have finally reached the *phenomenal forms* which the vulgar economist *starts out from*: land rent stemming from earth, profit (interest) from capital, wage from labor ... since these three (wage, land rent, profit [interest]) are the sources of income for three classes, of landowners, capitalists, and wage laborers, the final outcome is *class struggle* which will end this movement and all this shit'. (*Werke*, vol. 32, pp. 74f). [*Selected Correspondence*, p. 245 (adapted).]

[28] 'The recognition (*Erkennung*) of the products as its own, and the judgment that its separation from the conditions of its realization is improper – forcibly imposed – is an enormous [advance in] awareness [*Bewusstsein*] ...' Marx, *Grundrisse*, p. 463.

[29] Marx, *Grundrisse*, p. 494. When Marx's early *Philosophical and Economic Manuscripts* were published in the thirties, they became a real sensation and inspired a vast literature. The publication of *Grundrisse*, which contain preparatory work for *Capital*, from Marx's *mature* period of late 1850s, and which form an extraordinarily important

link between the *Manuscripts* and *Capital*, in turn passed virtually unnoticed. The significance of *Grundrisse* can hardly be exaggerated. They prove above all that Marx *never abandoned* the philosophical problematique, and that especially concepts of 'alienation', 'reification', 'totality', the subject—object relationship, etc., which certain ignorant Marxologists would be happy to declare as sins of Marx's youth, were parts of the *permanent* conceptual equipment of Marx's theory. Without them, *Capital* would be incomprehensible.

³⁰'All production is an objectification of the individual', Marx, *Grundrisse*, p. 226.

³¹'though bourgeois economists see how production works *within* capitalist relations, but do not see how these relations themselves are produced' *Marx—Engels—Archiv*, Moscow 1933, vol. 2, p. 176.

³²See Marx, *Contribution to a Critique of Political Economy*, New York 1970, p. 32.

³³See *Marx—Engels—Archiv*, p. 6.

³⁴'The functions that the capitalist carries out are but consciously and voluntarily performed functions of capital itself — of value valorizing itself by ingesting live labor. The capitalist functions only as personified capital, as capital as a person, just as worker is personified labor'. *Marx—Engels—Archiv*, p. 32.

³⁵'The concept of capital contains the capitalist', Marx, *Grundrisse*, p. 512.

³⁶Marx, *Grundrisse*, p. 712. (Emph. Kosík—Ed.)

³⁷The question, what is labor, is frequently answered with a *sociological* definition which characterizes it as 'all actions which man performs on matter for a practical purpose, aided by his brain, his hands, tools and machines, actions which in turn affect and modify man'. G. Friedmann, 'Qu'est-ce que le travail?' *Annales* 1960, no. 4, p. 685. Friedmann and Naville are two of the most important sociologists of work influenced by Marxism. We selected Friedmann's essay precisely as a representative example of the theoretical confusion with which justified demands for historical concreteness are intertwined with uncritical empiricism and sociologism. However, Friedmann's essays are a valuable contribution to his discipline, to the sociology of industry, technology and work.

³⁸The appropriate name for this apologetics is 'theology of work', and among its authors are not only Christian theologians. It is of course not accidental that Thomism has been paying great attention to the problem of labor. Modern Thomist authors (Vialatoux, Bartoli, Ruyer, Lacroix) direct their essays on work against materialism, which does not prevent them from taking over from Marxism, their opponent, its arsenal of facts.

³⁹The fourth chapter, 'Praxis and Totality', will elaborate this assertion further.

⁴⁰H. Marcuse, 'On the Philosophical Foundations of the Concept of Labor in Economics', *Telos* 16, summer 1973, p. 13. We shall have more to say about this important essay, whose best parts have yet to be improved upon.

⁴¹Although Sartre correctly states that the intellectual horizon of Marxism cannot be crossed in our epoch, he 'neglects' to add, *also* of Marxism as an 'ontology of man'. Cf. Sartre, *Search for a Method*, New York 1968, p. 30.

Sartre grounds his justification of existentialism (of existential philosophy and ontology) as an indispensible complement of Marxist philosophy precisely on this 'omission'.

⁴²'[T]he concept of labor appears [in Hegel] as a fundamental happening [*Grundgeschehen*] of human Dasein, as an abiding happening that constantly and continually spans the whole of man's *being* and at the same time involving even man's 'world'. Here labor is precisely *not* a specific human 'activity' . . . rather, labor is that in which every single activity is founded and to which they again return: a *doing* [*Tun*]'. H. Marcuse, op. cit., p. 13 (adapted). This important essay suffers from several main shortcomings: first, it does not distinguish between labor and praxis, which is an error that recurs

traditionally in most essays on praxis and labor: labor is characterized as the essence of praxis and praxis is defined essentially as labor; second, it does not distinguish between the philosophical and the economic concepts of labor, and thus it cannot objectively appreciate Marx's contribution; and third, it identifies objectivation with objectification, which renders the author vulnerable to subjectivism and introduces chaos and inconsistency into elaborating the problem of labor.

⁴³ Cf. esp. G. Lukács, *Der junge Hegel*, Berlin 1954, pp. 389–419. [*The Young Hegel*, London 1975, pp. 338–364.]

⁴⁴ Ivan Dubský treats the dialectical pairs of particular–general, subject–object and theory–praxis in Hegel's philosophy in his essay *Hegels Arbeitsbegriff und die idealistische Dialectik*, Prague 1961, pp. 30–44.

⁴⁵ In this sense, both beast and man are 'naturally' practical beings. See in this context Marx's polemic with Wagner, where he states that man does not 'stand' in reality but acts in it practically, in order to satisfy his needs.

⁴⁶ Linguistically, we feel it more appropriate to distinguish animal and human craving, instead of using the literal translations of Hegel's *Begierde* and *Trieb*.

⁴⁷ 'The animal exists only in the moment, it sees nothing beyond it; man lives in the past, in the present, and the future'. Diderot, *Oeuvres*, ed. Assézat, vol. 18, p. 179, as quoted in Poulet, *Human Time*, p. 187.

⁴⁸ Marx, *Capital*, vol. 1, p. 189. [The German text is closer to Kosík's argument: Die Arbeit 'sich aus der Form der Unruhe in die des Seins, aus der Form der Bewegung in die der Gegenständlichkeit' umsetzt.–Tr]

⁴⁹ That the problem of man's time is linked with his objective activity is a basic point in which materialist philosophy differs from the existential conception of temporality.

⁵⁰ Cf. Marcuse, op. cit., p. 23.

⁵¹ In this context we have to mention that A. Smetana, as opposed to Schelling, did not consider art to be a matter exclusively for the genius. In the spirit of his time he espoused a far more democratic conception of artistic creation. Smetana conceived of art in a broad and revolutionary–anticipatory sense as of the *free process of forming human conditions*. See A. Smetana, *Sebrané spisy* (Collected Works), vol. 1, Prague 1960, pp. 186f.

⁵² The relationship between necessity and freedom is an historically conditioned and an historically variable one. From the materialist perspective it is entirely consistent that Marx would link the problem of freedom with the creation of *free* time, an important moment of which is the shortening of labor time, and that in this sense he would translate the problem of necessity and freedom into terms of the relation of labor time and *free* time. 'But free time, *disposable time*, is wealth itself, partly for the enjoyment of the product, partly for the free activity which – unlike labor – is not dominated by the pressure of an extraneous purpose which must be fulfilled, and the fulfillment of which is regarded as a natural necessity or a social duty, according to one's inclinations'. K. Marx, *Theories of Surplus Value*, Moscow 1971, vol. 3, p. 257. The idea of free time as organized leisure is entirely foreign to Marx. Free time is not identical with leisure which can be a *part* of historical alienation. The existence of *free* time assumes not only the shortening of labor time but also the abolition of reification.

⁵³ Such is the case of romanticism and surrealism. Their defense leads to ill-considered conclusions, as evidenced by the following statement: 'The basis of the surrealist procedure is not Hegelian reason or Marxist labor; it is liberty'. F. Alquié, *The Philosophy of Surrealism*, Ann Arbor 1965, p. 83.

⁵⁴ This contradictory character of the historical process has been emphasized by Marx: 'The course of social development is by no means that because one individual has satisfied his needs he then proceeds to create a superfluity for himself; but rather because one individual or a class of individuals is forced to work more than required for the

satisfaction of its needs – because *surplus* labor is on one side, therefore not-labor and surplus wealth are posited on the other. In reality the development of wealth exists only in these opposites'. Marx, *Grundrisse*, p. 401.

[55] Marx, *Capital*, vol. 3, p. 359; cf. also *Grundrisse*, pp. 712, 609.

[56] Schelling, *Werke*, vol. 2, p. 622.

[57] 'Total life on the level of economics does indeed exist, but only in rather rare limited situations. We find ourselves precisely on the level of economics during crises (wars, famines, etc.) because what counts then is immediate life: eating, staying warm, etc'. R. Callois, 'Le monde vécu et l'histoire', *L'homme, le monde, l'histoire*, Paris 1948, p. 74.

[58] 'Political economy has to do with the specific social forms of wealth or rather of the production of wealth'. Marx, *Grundrisse*, p. 852.

CHAPTER IV

PRAXIS AND TOTALITY

Philosophical thinking of a given epoch will concentrate different aspects of its work in one central concept which will then appear in the history of philosophy as substance, *cogito*, Absolute Spirit, negativity, the thing-in-itself, etc. Without their philosophical problematique, these concepts would of course be empty. The historian who would sever solutions from problems would also transform the history of philosophy and of philosophical thinking into a senseless collection of petrified artifacts. He would turn the dramatic arena of truth into a wasteland of dead categories. Philosophy amounts above all to the posing of questions. It must therefore forever again substantiate its existence and its *raison d'être*. Every seminal discovery of the natural sciences, every great work of art changes not only the image of the world but especially man's very place in the world. The starting point of every philosophy is man's being in the world, the relation of man and cosmos. In everything he does, be it affirmative or negative, man always constitutes a certain mode of being in the world and determines (consciously or unconsciously) his position in the universe. Man establishes a relationship with the world through his very existence, and this relationship is already there before he ever starts contemplating it, before he turns it into an object of investigation, and before he practically or intellectually affirms or negates it.

PRAXIS

One important concept of modern materialist philosophy is that of praxis. Everyone knows what praxis is and what it is not, even without philosophy. Why then did philosophy turn this self-evident thing into a key concept? Or did perhaps praxis have to become a *philosophical* concept before it could dispel the semblance of certainty with which naive consciousness is always well informed in advance about praxis and practicality, about the relationship of praxis and theory, about practicing and practicism? Naive consciousness finds philosophy to be a world turned upside down — and rightly so: for philosophy does indeed turn *that particular* world upside

down. Philosophical *questioning* shatters the certainties of ordinary consciousness and of everyday fetishised reality when it questions their
appropriateness and 'reasonableness'. This is not to say that naive
consciousness is out of touch with philosophy, or that it is indifferent to
philosophy's results. Everyday consciousness appropriates results of philosophy and considers them its own. However, because this consciousness did
not undertake the *journey* of philosophy, and reached the latter's
conclusions effortlessly, it does not take them too seriously and instead
treats them as self-evident matters. That which philosophy exposed in
concealment, oblivion and mystification as being *evident*, is appropriated by
ordinary consciousness as *self-evident*. All that philosophy *has made* visible,
conspicuous and tangible sinks in this self-evidence back into anonymity
and non-evidence.

The only part of the great discovery of materialist philosophy that
uncritical reasoning preserved was the idea that praxis is something
immensely important and that the unity of theory and praxis holds as a
supreme postulate. But the *original philosophical questioning*, in whose light
praxis had been discovered, disappeared, and the idea preserved merely the
importance of the principle. Consequently, the content of the concept of
praxis *changed*, and the unity of theory and praxis came to be realized and
grasped in different epochs in quite peculiar ways. In our analysis of labor
we pointed out one of the historical changes that has affected the concept
of praxis: praxis grasped as 'socialness', and Marxist philosophy as the
teaching about the 'socialness of man'. In another transformation, 'praxis'
turned into a mere category and functioned as a correlate of cognition and
as a fundamental concept of epistemology. In other metamorphoses, praxis
was identified with technology in the broad sense of the word, and
conceived and practiced as manipulation, as a technique of conduct, as the
art of disposing of people and things, in short, as the power to manipulate
and as mastery over material both human and inert. Modifications in
comprehending and practicing praxis correlated with corresponding
changes in the concept, task and sense of philosophy, and in the concept of
man, world, and truth.

In what sense, and in what philosophical tradition has materialist
philosophy hoisted praxis as *its* central concept? At first glance it might
seem (and this impression has frequently 'materialized' in particular
opinions) that a generally known reality and a trivial self-evident matter was
given a *philosophical significance* and was *generalized*: after all, had not
thinkers and practitioners of all times known man as practically active? Is

not all of modern philosophy (in a conscious distinction from medieval scholastics) formed as the knowledge and cognition by which we are to make ourselves the 'masters and possessors of nature'?[1] And had not classical philosophy of history (Vico, Kant, Hegel) already expressed the thought that people act in history, and that their actions lead to consequences and results they had not intended? Has materialist philosophy perhaps merely gathered the scattered and isolated findings of previous eras concerning praxis as the action of people, as industry and experiment, as the historical cunning of reason, and then synthesized them into a basis for a scientific interpretation of society? Similar considerations would merely lead us over another path back to the opinion that in Marxism, philosophy has been abolished and translated into a dialectical theory of society, or in other words, that praxis is not a philosophical concept but a category of a dialectical theory of society.

The problem of praxis in materialist philosophy cannot be explained from the relationship of *theória* and *praxis* or of contemplation and activity, whether with emphasis on the primacy of theory or contemplation (Aristotle and medieval theology), or conversely of praxis and activity (Bacon, Descartes and modern natural sciences). Emphasis on the primacy of praxis over theory goes hand in hand with devaluing the *significance* of theory, which in relation to praxis is degraded to *mere* theory, and at the same time, the sense and *content* of praxis are grasped just as poorly as when antiquity emphasized the primacy of theory. The primacy of praxis over theory, which turns up in formulations such as that knowledge is power, or in substantiating the importance of theory *for* praxis,[2] stems from a *particular historical* form of praxis in which the *essence* of praxis both manifests and conceals itself in a characteristic fashion.

The secularization of nature and the discovery of nature as a conglomeration of mechanical forces, as an object of exploitation and subjugation, goes hand in hand with the secularization of man, who is discovered as a being that can be molded and formed or, translated into an appropriate language, that can be manipulated. Only in this connection can one grasp the historical significance of Machiavelli and the sense of Machiavellism. The naive journalistic view judges Machiavellism through the prism of contemporary manners of ruling, and considers him a guide to the politics of trickery and deceit, of the dagger and poison. Machiavelli, however, was not merely an empirical observer or a talented commentator of historical texts who would have written up the current practices of Renaissance princes and traditional events of the Roman world, and *generalized* them all.

His place in intellectual history is rather one of a penetrating analyst of human reality. His basic discovery, corresponding to Bacon's operational science and to the modern conception of nature, is his concept of man as a disposable and a manipulable being.[3] Scientism and Machiavellism are two facets of the same reality. This is the basis for formulating a conception of politics as a calculable and rational technique, as scientifically predictable manipulation with human material. It is unimportant for this conception and for the 'praxis' that corresponds to it whether man is by nature good or evil: good or evil, his nature is always *malleable* and he can therefore be the object of calculable and scientifically-based manipulation. Praxis arises in the historical form of manipulation and procuring or, as Marx was to prove later, in the form of a dirty haggler.

From the practical perspective, and from the perspective of praxis as manipulation, procuring and disposition, one can be either an apologist or a critic of 'praxis'. This affirmative or negative attitude is, however, limited to the sphere of the pseudoconcrete and can therefore *never* uncover the real character of praxis. Nor can the character of praxis be inferred from the distinction between the man of praxis and the man of theory or between practicality and theorizing, for this distinction is itself based on a particular *form* or *image* of praxis and indicates this particular form only, rather than praxis as such.

The problem of praxis in materialist *philosophy* is not based on distinguishing two areas of human activity or on a typology of possible universal intentionalities of man,[4] nor does it stem from an historical form of a practical relationship with nature and with man in which both would be objects of manipulation. Rather, it is formed as a philosophical answer to a philosophical question: *Who is man, what is socio—human reality, and how is this reality formed?*

In the concept of praxis, socio—human reality is discovered as the opposite of givenness, i.e. at once as the process of forming human *being* and as its specific form. *Praxis is the sphere of human being.* In this sense, the concept of praxis is the outcome of *modern* philosophy which has emphasized, in a polemic against the Platonic—Aristotelian tradition, the authentic character of man's creating, as of an ontological reality. Not only are existents 'enriched' by man's work, but his work is where reality indeed manifests itself in a particular way and where access to it is negotiated.

Something essential happens in man's praxis, something that contains its own truth in itself rather than merely pointing elsewhere, something that is also of ontological importance.[5]

In its essence and generality, praxis is the exposure of the mystery of man as an onto-formative being, as a being that *forms* the (socio–human) reality and *therefore* also grasps and interprets it (i.e. reality both human and extra-human, reality in its totality). Man's praxis is not practical activity as opposed to theorizing; it is the determination of human being as the process of *forming* reality.

Praxis is active and self-producing in history, i.e. it is a constantly renewing, practically constituting unity of man and world, matter and spirit, subject and object, products and productivity. Since the socio–human reality is *formed* by praxis, history becomes a practical happening in which what is human is distinguished from what is inhuman. What is human and non-human is not preordained; it is determined in history through a process of practical discrimination.

In the preceding chapter we pointed out the lack of conceptual clarity in delimiting praxis and labor: others have defined labor as praxis and characterized praxis by reducing it to labor.

Since praxis is a specific mode of man's being, it permeates the essence of his being in all its manifestations, rather than determining only some of its aspects or traits. Praxis permeates the *whole* of man and determines him in his totality. Praxis is not an *external* determination of man: neither a machine nor a dog have or know praxis. Neither a machine nor a dog know the fear of death, the anxiety of nothingness, or the joy of beauty. Man does not build a culture and a civilization, his socio–human reality, as a shield from mortality and finitude, but discovers his mortality and finitude *only* on the basis of civilization, i.e. on the basis of his objectification. How did that break ever come about, in which the animal-man that had known nothing of death or of mortality and therefore had known no fear *of* death either, was transformed into the animal-man who recognized death as the outcome of his future and has consequently ever since been living under the sign of death? According to Hegel, this break occurred in the struggle for recognition, in the battle for life and death. This struggle, however, could have taken place only if man had already discovered the future as a dimension of his being, which is possible *only* on the basis of labor, i.e. of the objectification of man. The struggle for life and death *may not* end in death, both fighters must remain alive, although they both do wager their life. This *premise* of the master–slave dialectics is, however, an historical prerequisite. In the struggle for life and death, man lets the other one live, and the other opts for slavery rather than for death, *only because* they *both* know about the future and know what awaits them: *either* mastery *or* slavery.[6]

He who prefers slavery to death, and he who wagers his life *in order* to be recognized as man-the-master, are both men who *already* know time. Man surrenders to his (future) fate of a slave or fights for his (future) position of a master only because he chooses his present from the perspective of the future, and thus forms his *present* on basis of a project of a future. Both men form their present and their future on the basis of something that is not yet is.

The future is known to man only in its immediacy. The slave becomes a slave with a slave's consciousness which at first is devoid of any hope or supposition that slavery ever will or might end: he enters his future as he would eternity, forever. Similarly, the master. It takes the dialectics of the actual course of affairs to *transform* the future, to invalidate the *immediate* future as untrue or one-sided, and to elevate a *mediated* future as the truth: in the dialectics of the master and the slave, slavery is the only passable path and the only way to freedom, whereas mastery proves to be a dead end. But, how does man know even about his immediate future, enough to undertake the struggle for recognition? The three-dimensionality of time as a form of his own being manifests itself to man and constitutes itself in the process of objectification, i.e. of labor.

Thus apart from the moment of *labor*, praxis also includes an *existential* moment: it manifests itself both in man's objective activity by which he transforms nature and chisels human meanings into natural material, and in the *process of forming* the human subject in which existential moments such as anxiety, nausea, fear, joy, laughter, hope, etc., stand out not as positive 'experiencing,' but as a part of the struggle for recognition, i.e. of the process of realizing human freedom. Without the existential moment, labor would cease to be a component of praxis. Man frees himself through slave labor only providing that:

(1) his labor develops as the labor of slaves and not as the labor of an isolated slave, and thus allows for the potentiality of slave solidarity;

(2) slave labor has its counterpart in the master's not-labor, and is really incorporated in the master—slave social relation; only in this *practical relationship* does there exist a possibility for comparison, and thus for recognizing the profound differences in position and life;

(3) the labor of the slave is felt and conceived of as slave labor, and exists as such in the slave's consciousness. *This* consciousness has a tremendous revolutionary potential. A mere objective relationship to nature cannot generate freedom. What in certain historical phases appears as the 'impersonality' or the 'objectivity' of praxis, and what false consciousness

elevates as the most proper practicality of praxis, is praxis only as manipulation and procuring, i.e. praxis in its fetishized form. Without the existential moment, i.e. without the struggle for recognition which permeates man's entire being, 'praxis' sinks to the level of technique and manipulation.

Praxis is both the objectification of man and the mastering of nature, and the realization of human freedom.[7]

There is yet another dimension to praxis. Though it is a specific human reality that is formed in the happening of praxis, reality that is independent of man exists in it *in a certain way* as well. In praxis, man's *openness* toward reality in general is formed. The onto-formative process of human praxis is the basis for the possibility of ontology, i.e. for understanding being. The process of forming a (socio–human) reality is a prerequisite for disclosing and comprehending reality in general. Praxis as the process of forming human reality is also a process of uncovering the universe and reality in their being.[8] Praxis is not man's being walled in the idol of socialness and of social subjectivity, but his openness toward reality and being.

All manner of theories of social subjectivism (sociology of knowledge, anthropologism, philosophy of care) have *walled* man in a subjectively conceived socialness and practicality: in their opinion, man *expresses only* himself and his social position in his creations, and projects his subjectively objective situation into forms of objectivity (science). By contrast, materialist philosophy believes that on the basis of praxis as an onto-formative process man also develops his historical ability to reach out beyond and outside himself, to be disclosed to being in general. Man is not walled in his animality or in his socialness, since he is not an anthropological being. Rather, he is disclosed to the understanding of being on the basis of his praxis. Consequently, he is an anthropo-cosmic being. Praxis has been discovered as the foundation of a real active center,[9] as the real historical mediation of spirit and matter, culture and nature, man and the universe, theory and action, existents and existence, epistemology and ontology.

We learn about the world, things and processes only so long as we 'form' them, i.e. so long as we spiritually and intellectually reproduce them. This spiritual reproduction of reality cannot be grasped other than as *one* possible practical human relationship with reality, as the one whose most fundamental dimension is the *process of forming* (socio–human) reality. Without the process of forming socio–human reality, without producing it, it would be impossible to spiritually and intellectually *reproduce* it.

How is it at all possible to understand reality? How can one understand the relationship between the finite cognitive being and the rest of the world? Man can understand things and their being, the world in its particularities and in its totality only on the basis of his openness that develops in praxis. In praxis and on the basis of praxis, man transcends the closed character of animality and of inorganic nature and constitutes his relationship with the world in its totality. In his openness, man as a finite being transcends his finitude and establishes contact with the totality of the world. Man is not only a part of world's totality: without man and without his cognition as parts of reality, reality and its cognition would remain incomplete. But movements of the world's totality include both the way in which this totality uncovers itself to man, and man in uncovering this totality.[10]

The totality of the world includes man, with his relation of a finite being to infinity, and with his openness toward being. Upon this is based the very possibility of language and poetry, of questioning and knowing.

HISTORY AND FREEDOM

Before we can argue what history is *like*, we have to know what history *is* and how it is possible. Whether history is absurd and cruel, tragic or farcical, whether a plan of providence or an immanent law is realized in it, whether it is the arena of licence and hazard or the field of determinism – all these questions can be satisfactorily answered only if we know what history is.

The historian investigates what has happened in history while the philosopher asks what history is and how it is indeed possible. The historian deals with the history of the Middle Ages or of modern times, of music or of painting, of ideas or of celebrities, with the history of a nation or of the whole of mankind. The philosopher, in turn, wants to know what are the suppositions of any history and how can anything like history exist at all. His questions do not impinge upon the specialized problematique of the historian, but they inquire into the *presuppositions* of that discipline, doing work that the historian could not accomplish with his tools and within his discipline.

Man had been creating history long before he recognized that he is an historical being, and has been living in it ever since. But the historical consciousness that has discovered history as an essential dimension of human reality does not yet testify, in and of itself, to the *truth* of what history is.

Every *profound* attempt to formulate the specificity of history features some mystification, and this is true also of classical historicism, from Vico to Hegel. It is as though the profound insights were internally connected with mystification. Nineteenth-century positivist and evolutionist trends deleted Hegelian speculation and mystification from history but in so doing they impoverished it, even as they burdened it with their own *new*, vulgar mystifications. Can the depth and the multidimensionality of history be understood without falling into mystifications? That depends on how we explain its character and function. What is the role of providence in Vico's, Schelling's or Hegel's philosophy of history? Is it merely a religious and theological element or does it play yet another role in their philosophy, a role independent of its religious provenance? Is the philosopher who introduces providence into his concept of history a religious thinker, or is there a *definite* reason that would compel even a non-religious thinker to employ 'providence' as a constructive element of history? To pose the question in this way assumes either that the religious problematique is taken to be nonsense and deception, or that modern history, including modern intellectual history, is viewed as an extensive process of secularizing the Christian-theological world view. However, the matter looks *entirely* different if we consider religious problems to be a *mystified* expression of real problems: in which case modern intellectual history will no longer appear as an extensive process of secularization, and will instead show up in its true form, as an attempt to *rationally* solve problems which religion had expresssed in a *mystified manner*. From this perspective, the motivation for introducing providence into history is secondary.

Historical providence comes under different names, but the problem remains the same: without providence, without the 'invisible hand,'[11] without the 'cunning of reason,' or the 'intention of nature,'[12] history would be incomprehensible: it would appear as the *chaos* of discrete acts of individuals, classes and nations, as eternal *change* condemning every work of man to extinction, as the *alternation* of good and evil, of humanity and inhumanity, of positive and negative, with no guarantee that good and humanity would eventually *have* to prevail in this struggle. *Providence is the grounds for and the guarantee of history's reasonableness.* The 'cunning of reason,' the 'intention of nature,' or the wisdom of the 'invisible hand' are not metaphors adorning the trivial fact that the real result of conflicting individual interests differs from what people had originally intended, i.e. that the result of human action does not coincide with its intentions. The classical philosophy of history postulates that the result of disharmony between the intentions and the results of human actions be a *reasonable* reality. The chaotic and unpredictable conflict of human actions and the disharmony between the necessity and the freedom of human activity, between what people intend and what they actually do, between who they think they are and who they are actually, all this gives rise to something that people had not anticipated or intended, but what is, nevertheless, *reasonable*. *If* people were left to their own devices, to their passions and interests, to their egotistic industry and particular prejudices, history would not progress to an eschatological culmination but would go on and on as the eternal and senseless circulation of reason and unreason, good and evil, humanity and inhumanity; it would indeed be a 'system of godlessness and atheism.' If history is reasonable and has sense, it is only because a *higher* intention, reason, or plan of providence is manifest and realized in it. 'History as a whole is a gradual, step-by-step revelation of the Absolute.'[13] Acts of man do not have sense and reason in and of themselves but *acquire* such sense and reason *with respect to* the plan and reason of providence. This conception has two important implications: according to it, history *is formed* as a *dialectical* process, but people are mere *instruments* of the dialectics of history. The unity of necessity and freedom is realized in history, but freedom is in the last analysis only fictitious, and so is consequently the *unity* of necessity and freedom. This contradiction shows the greatness and the limitations of the classical conception of history.[14] Classical philosophy had correctly *formulated* the problem of history but did not resolve it. More precisely: it abandoned the correct *original* formulation *in the course* of seeking a solution to it. The original formulation was this: Neither absolute law, nor absolute freedom reign supreme in history, there is nothing absolutely necessary or absolutely accidental in it; history *is* the dialectics of freedom and necessity. The *solution* is suggested by well-known statements: freedom is *recognized* necessity, freedom is a figment.[15] For history to be reasonable and to have sense, it has to be designed in a plan of providence in which historical individuals (outstanding personalities, nations, classes) act as conscious or unconscious agents of preordained necessity. People *act* in history, but only *seemingly* do they make history: history is the realization of necessity (the plan of providence, a foreordained harmony), and historical personages are its tools and executive arms.

In the 20th century it is no longer a great scientific discovery to expose this concept of history as a mystification and to criticize it as the 'religion of freedom' or as 'romanticism.' In philosophy of history, the fate of man is indeed infallibly guaranteed by an infinite force which may have different names (Humanity, Reason, the Absolute, Spirit, Providence), but has always the same task: to overcome the defects, correct the deviations, and lead to the definitive triumph of good. Philosophy of history is indeed based on the assumption that the ultimate success of human activity is necessarily guaranteed by the metaphysical structure of the world.[16] But ever since Marx found that history does exactly nothing, and that everything in it, including history itself, is the doing of man, the primary task has been not to list the shortcomings of philosophy of history, but to examine the *causes* of its fundamental mystification. History is made by people. But why does it *seem* that people are mere agents or executors of this 'making of history'? People act in history at their own risk and danger. But why do they act in the belief that they have been summoned by a higher power to perform historical deeds? History is a product of mankind. But why do people over and over again act *as though* they were the agents or trustees of this product? The individual gathers his *courage* for action, *justifies* and *substantiates* his action by transforming himself, as it were, into an agent of a transcendental power and by turning into the spokesman, deputy and regent of God, Truth, Humanity. He does not realize *his own* interests but carries out the iron laws of History. From the point of view of technique and performance, killing a man is a simple matter.[17] The dagger, sword, axe, machine gun, pistols and bombs are effective and well-tested tools. But the 'simple matter' immediately becomes complicated if we shift from 'performance' to 'evaluation,' from 'technology' to 'society.' He who kills for his personal motives, privately and on his own accord, is a *murderer*. He who kills with higher authorization and 'in the interest of society' is *not* a murderer. If the perpetrator of the act is an instrument of his own intention or of his passion. he commits a crime. If he is a *mere* instrument, it need not be a crime. If I were to kill a man in and of myself, I might get scared of my own action, back away, and not carry out my intention: there is nothing cowardly and dishonorable in refraining from *this* action. But were I to kill with 'higher authorization,' by order of the Nation, Church, or Historical Necessity, I could not refrain from 'my' action, lest I be branded a coward. My act is not murder but revenge, trial, execution of justice, civic duty, an heroic deed. But the 'truth' of history, i.e. its concreteness, multidimensionality and reality, is such that a particular act can be at once murderous and heroic, that murder can be elevated to heroism and heroism degraded to murder, that particular interests can be declared general interests, and real general interests debased as individual intentions.[18]

History 'includes' both heroism and crimes. Burning heretics at the stake is not an 'excess' of the times, an anomaly or an abnormality of an 'unenlightened epoch,' and thus historically a peripheral matter; rather, it is as normal and constitutive a component of feudalism as is papal infallibility and serf labor. Philosophy of history has appreciated the role of evil as a constitutive element in the process of forming socio–human reality, but in the overall metaphysical construction of the world this role was preordained as well: evil is a *component* of good; its *positive* role is in preparing and evoking good; with respect to the ultimate triumph of good, guaranteed by metaphysical necessity, evil plays a positive role too.

And yet: if the metaphysical constitution of the world which generates the victory of good, gives history its sense and lays down the reason of history were not the immanent structure of reality but only one of the *historical images* of the world; if history were not preordained and if there were no cosmic signs from which man might divine that the victory of good in history is guaranteed once and for all, and absolutely; if the Reason through which Hegel contemplated history, so that it be reasonable, were not the 'unbiased' and the transhistorical reason of the objective observer but the dialectically formulated reason of the Christian-theological world view; if all this, would it then follow that history is absurd and senseless, that history and reason exclude one another? The critique of philosophy of history implies above all that a providentially constructed reason does not allow for a rational grasp of history. Providential reason has *designed* history as reasonable *in advance*, and only on the basis of this unsubstantiated metaphysical assumption have the concepts of the 'cunning of reason,' the 'invisible hand,' or of the 'intention of nature' been constructed. Only thanks to them – i.e., in a mystical dialectical metamorphosis – does chaotic and particular human activity lead to a *reasonable* conclusion. History is reasonable only because it has been designed and ordained as reasonable *in advance*. Related to this reason, all un-reason, evil and negativity, victims and suffering, all these become a negligible magnitude or a secondary effect. Not even in Hegel's conception is historical reason dialecticized consistently. Consistent dialectization of historical reason requires the abolition of the metaphysical–providential foundation of this reason. Reason is not laid down throughout history ahead of time, in order to be revealed as reason in the historical process, but rather it *forms itself* as reason in the course of history. According to the providential conception, reason *designs* history, and is itself gradually revealed in history's realization. By contrast, according to the materialist conception, only in history is reason first formed; history is not *reasonably preordained* but only *becomes* reasonable. Reason in history is not the providential reason of foreordained harmony and the metaphysically preordained triumph of good. Rather, it is the *conflict-reason* of historical dialectics in which *reasonableness* is the object of struggle, and every historical phase of reason is realized in conflict with historical unreason. In history, reason becomes reason as it realizes itself. There exists no ready-made, transhistorical reason that would reveal itself in historical events. Historical reason arrives at its reasonableness through realization.

What does man realize in history? The progress of freedom? The plan of providence? The course of necessity? In history, man realizes himself. Before history and independent of history man not only *did not know* who he was; only in history *is he* even a man at all. Man realizes himself, i.e. humanizes himself, in history. The span of this realization is so tremendous that man characterizes his own performance as *inhuman*, though he knows well that only man can act inhumanly. Once the Renaissance discovered that man is his own creator and can cast himself into whatever he chooses, be it an angel or a wild beast, a human lion or a human bear, or indeed *anything else*,[19] it soon became obvious that human history is the unfolding of these 'possibilities' over time. The sense of history is in history: in history, man explicates himself, and this historical explication, amounting to the process of forming man and humanity, is history's only sense.[20]

In history it is man and only man who is realized. Therefore history is not tragic — though there is the tragic in history; it is not absurd, though the absurd does develop in history; it is not cruel, though cruelties are perpetrated in history; it is not ridiculous, though comedies are acted out in history. In history, individual epochs follow one another in a certain order and in a law-like manner, but they never lead to a *definitive* culmination or to an apocalyptic end. No epoch in history is *nothing but* a transition to some other stage, just as no epoch towers over history as a whole. Every epoch is a conjunction of the three-dimensionality of time: its preconditions are rooted in the past, its consequences reach into the future, and its structure is anchored in the present.

The first basic premise of history is that it is *created by man*, but its second, equally basic premise is the necessity for *continuity* of this creation. History is only possible at all because man does not always start over again from the beginning and instead follows up the road and results of past generations. If mankind were to start each time from square one and if every action were without suppositions, mankind would never budge from one place and its existence would move in a circle of periodic recurrence of an absolute beginning and an absolute end.

The interconnection of objectified and objectivised praxis of mankind, labelled as substance, objective spirit, culture or civilization, and decoded in materialist theory as the unity of production forces and production relations, forms the historically attained 'reason' of society, which is independent of any particular individual and is thus transindividual, but which really exists *only through* the activity and reason of *individuals*. The objective social substance, in the form of materialized production forces, language and forms of thinking, is *independent* of the will and consciousness of individuals, but it *exists* only through *their* activity, thinking and language. Machines that are not set into motion by human action, languages that people do not speak, logical forms in which people do not express their thinking are *either* dead props, *or* sheer nonsense. Objectified and objectivised praxis of mankind, in the form of production forces, language, forms of thought, etc., exists as the *continuity of history only* in connection with the *activity* of people. The objectified and objectivised praxis of mankind is the lasting and fixed element of human reality. In *this form* it resembles a reality more real than praxis or any human activity. This is the basis for the *possibility* of inverting the subject into the object, i.e. for the *fundamental* form of historical mystification.[21] Since objectified and objectivised praxis of man *survives* every individual and is independent of him, man interprets himself, his history and his future first and foremost from his own *creations*. Compared with the finitude of an individual life, objectified and objectivised praxis of mankind embodies the *eternity* of man. Compared with the hazards and fragility of individual existence, the 'social substance' represents permanence and the absolute. Compared with the limited reason and the unreasonableness of the empirical individual, this substance amounts to real reason. When man considers himself a tool or a spokesman of providence, of the absolute spirit, History, etc., i.e. of an absolute force that infinitely transcends his own possibilities and reason, he falls into mystification. This mystification is, however, not a rational expression of nonsense, but a mystified expression of a rational reality: objectified and objectivised praxis of mankind enters people's heads as a metaphysical being independent of mankind. Man creates his eternity only in an objectified, i.e. in an historical, praxis and in its creations. In an alienating inversion, the objectified and objectivised praxis of mankind turns into the mystical *subject* in which man seeks a guarantee against chance, unreason and the fragility of his own individual existence.

People enter conditions independently of their consciousness and of their will but 'once there,' they transform these conditions. Conditions do not exist without people, or people without conditions. This is the basis for the development of a *dialectic* between *conditions* that are *given* for every individual, for every generation, epoch and class, and action that unfolds on the basis of ready-made and given prerequisites.[22] Conditions stand out as prerequisites of this action; the action in turn invests them with a particular *sense*. Man transcends conditions not *primarily* in his consciousness and intentions, in his ideal project, but in his praxis. Reality is not a system of my meanings nor is it transformed in accordance with the meanings my project gives it. It is in his action that man inscribes meanings into the world and forms a structure of meanings in it. In my project, my fantasy and imagination, in my dreams, I can *transform* the four walls into which I have been thrown in chains into a kingdom or into a realm of freedom; but these ideal projects will not make the four walls any less a prison, and my confinement within them will not be any less unfree. For the peasant serf, 'conditions' is the immediate natural situation of life; indirectly, through his action, resistance or in a peasant uprising, he gives them the signification of a prison: conditions are *more* than just conditions and the peasant serf is *more* than a part of conditions. Conditions and man are constitutive elements of praxis which is in turn the fundamental prerequisite for transcending conditions. The situation of human life turns into unbearable and inhuman conditions *with respect* to the praxis that is to transform it. People act under certain conditions and their practical action gives conditions a meaning. The forms of social *movement* turn into *fetters*. Social orders, formations, forms of coexistence are the *space* in which social movement is realized. In a certain situation, this space becomes limited and is felt as bondage and unfreedom. Starting with Hobbes, the materialist tradition has determined freedom as the space in which an object moves. From a mechanistically conceived space, which is independent of the movement and the character of the object, and which forms only the outside delimitations of the object's movement, the materialist conception has progressed to the French Enlightenment's theory of social environment, and has culminated in the view that freedom is an historical process, expanded and realized by the activity of an 'historical body,' i.e. a society, class, individual. Freedom is not a state, but rather an historical activity that forms corresponding modes of human coexistence, i.e. a social space.

MAN

Gods exist only for those who recognize them. Outside the country's border they become a piece of wood, just as the king becomes a commoner. Why?

Because a god is not a piece of wood but a social relation and product. The critique with which Enlightenment took religion away from people and argued that altars, gods, saints and temples are 'nothing but' so much wood and canvas and stone was philosophically inferior to the creed of the believers — for gods, saints and cathedrals most certainly are not just so much wax and canvas and stone. They are a social product, not a natural one, and nature can therefore neither create them nor substitute them. This naturalist conception created a distorted idea of social reality, of human consciousness and of nature. It understood human consciousness exclusively as the biological function of the organism's adaptation and of its orientation in an environment, characterized by two basic elements: impulse and reaction. While in this way one might explain consciousness as a property common to all higher animals, one will not explain the specificity of human consciousness. Human consciousness is the activity of the subject who forms a socio-human reality as a unity of being and meanings, of reality and sense. While traditional materialism emphasized the material character of the world, transcendentalism emphasized the autonomy of reason and spirit as the activity of the subject. Its material character was separated from activity, because values and meanings are not inscribed in nature, and human freedom cannot be derived from a causal chain progressing from lichens and protozoa all the way up to man. While idealism insulated meanings from material reality and transformed them into an independent reality, materialist positivism on the other hand deprived reality of meanings. This completed the task of mystification, because the more perfectly man and human meanings would be eliminated from reality, the more real would this reality be considered.

But 'human reality' does not cease to exist even when cast out of science and philosophy. Otherwise we could not explain the periodically appearing waves of 'anthropologism' which draw attention to the problem of the 'forgotten' man.

It has been suggested that while man busies himself with everything possible between heaven and earth, he neglects himself. A typology has been elaborated which claims to prove that only periods of man's isolation are propitious for philosophical anthropology, i.e. to the cognition of man, whereas extroverted epochs deal with man in the third person, just as with rocks and animals,[23] and disregard his specificity. The need and the call for a philosophical anthropology is argued by suggesting that in no other historical epoch has man been so much of a problem to himself as he is now, when he has accumulated incomparably more knowledge concerning himself than ever before, but is also less sure than ever in the past of this knowledge.[24] And at the very time of 'anthropology's' culmination, there surfaces the opinion that 'anthropology' is not first and foremost a science of man (incidentally, a problematical science and one difficult to define) but rather a *'fundamental tendency'* of a time that has made man problematic.[25]

If then 'philosophical anthropology' wants to be a science of man and to study his place in the universe, the question which emerges first is this: Why is man *more* a man in *isolation*, when he deals with himself, than in 'extrovertness', when he investigates everything possible 'between heaven and earth'? Perhaps 'philosophical anthropology' emphasizes epochs of homelessness, isolation, and problematization of man because it has already *interpreted* the problem of man in a *definite* way, and considers only *certain* aspects of man as constituting a problem for anthropology?

In his orientation toward the outside world and in his investigation of natural laws man is no less a man than in his dramatic questioning of himself: *Quid ergo sum, Deus meus, quae natura mea*? If 'philosophical anthropology' privileges certain aspects and problems, it demonstrates that it has evolved not as the questioning of man's being and of his place in the universe, but as a reaction to a particular *historical situation* of people in the 20th century.

Philosophical anthropology strives to be a philosophy of man and to establish man as the basic problem of philosophy. Is this a justified pretension? Let us first of all suggest that the name 'philosophy of man' has several meanings. Philosophical problems are not inscribed in the universe but are formulated by man. What 'philosophy of man' means above all is that philosophical problems are formulated only by man, that only he philosophizes. Philosophy is one of man's activities. In this sense, *every* philosophy is a philosophy of man, and emphasizing the human character of philosophy by a specific attribute is superfluous.

But the 'philosophy of man' has yet a second meaning: *all* philosophical problems are essentially problems for anthropology, because man anthropologizes *everything* with which he is in practical or theoretical contact. All questions and answers, all doubts and findings testify first and foremost of man. In all his doing, from practical preoccupation to the investigation of trajectories of heavenly bodies, man above all defines himself.

'Philosophical anthropology' refers to Kant's famous questions:

(1) What can I know?

(2) What ought I do?

(3) What may I hope?

Kant adds a fourth question to these three: Who is man? The first question is answered by metaphysics, the second by morals, the third by religion, and the fourth by anthropology. But Kant explicitly notes that the first three questions can actually also be classified under anthropology since all three are related to the last question.[26] Who is that being which is asking what he can know, what he ought do, and what he may hope?

Depending on *where* one puts the emphasis, Kant's questions can be interpreted in the sense of finitude in man (Heidegger) or in the sense of man's share in infinity (Buber). But irrespective of the interpretation, the first three questions *predetermine* the fourth. Man is a being which learns what it can know, which learns what it ought do, which learns what it may hope. The first three questions define man as a *cognitive* subject and as the subject of cognition. Further generations have added to and improved upon this intellectual horizon, and have reached the conclusion that man is not only a cognitive being but also an experiencing and an acting being: man is the subject of cognition, the subject of experiencing, the subject of action. Thinking out this outline consistently, the world appears as man's project: the world is here only insofar as man exists.

In this second meaning, the 'philosophy of man' expresses the perspective of human subjectivity: the foundation and the point of departure for philosophy is not man, man in general, but a certain *conception* of man. Philosophical anthropology is a philosophy of man inasmuch as it conceives of man as of subjectivity.

The philosophy of man has, however, yet another, third, meaning. It is a programmatic discipline, which is to deal with neglected issues such as individual responsibility, the sense of life, the conflicting character of morality, etc. Philosophy of man is a name for the forgotten and the ignored, for the forbidden and the neglected. It is considered an indispensable *complement* which has to be added to philosophy as it stands, in order to update it and to have it provide answers to all questions. Leaving aside the elementary fact that it merely confers an ostentatious title on problems of ethics, the programmatic concept of a 'philosophy of man' suffers from an unbridgeable internal contradiction. The need for a 'philosophy of man' as a *complement* of philosophy reveals the obfuscation and the problem-ridden character of the basic principles of the very philosophy that clamors for an 'anthropological complement'. The basic design and matrix of this philosophy has either *left out* man entirely, or has included him only after transforming him into a non-man, i.e. after reducing him to a mathematical–physical magnitude. Now, under the impression of *outside* necessity, this philosophy feels the need to be supplemented with whatever it lacks – namely with man. A philosophy of reality without man is thus complemented with none other than a philosophy of man. We have two extremes here: on the one hand a concept according to which reality is a *reality of man*, and the world is a human project; on the other hand a concept according to which the world is authentic and objective only insofar as it is designed as world *without* man. This latter world is however *not* the authentic reality, but only one of the designs of human subjectivity, one of the possible ways in which man appropriates (and spiritually reproduces) the world. The physical image of the world, realized in modern natural sciences from Galilei through Einstein, is but one of the possible practical–spiritual approaches to reality: one of the ways to theoretically design (to spiritually reproduce) and to practically master reality. If this image is ontological (which is out of the question for materialist philosophy which grasps cognition as the spiritual reproduction of reality), i.e. if it is considered to be reality itself, and man is to search for his relation to and for his place in this 'reality,' he will manage to succeed only if he either transforms himself into a mathematical–physical magnitude, i.e. into a calculable component of an organized system, or if he arrays himself and counts himself in with such a system as its subject, i.e. as a theoretician, a physicist, a mathematician.

Without man, reality is not authentic, just as it is not (only) a reality of man. Reality is a reality of nature as the absolute totality, independent of man's consciousness but also of his existence. It is a reality of man who as one of nature's components forms in nature a socio-human reality that transcends nature, and who through history defines his place in the universe. Man does not live in two different spheres, nor does he inhabit history with one part of his being and nature with his other part. *Man is at all times at once in nature and in history.* As an historical, and thus as a social being, he humanizes nature but also knows it and recognizes it as the absolute totality, as the self-sufficient *causa sui*, as a precondition and prerequisite of humanization. In the cosmological concepts of Heraclitus and Spinoza, man recognized nature as the absolute and inexhaustible totality to which he forever anew defines his relationship, throughout history: by mastering the forces of nature, by learning the laws of natural events, in myths, poetry, etc. But regardless of the variability of man's approach to nature, of all progress in his mastery and knowledge of natural processes, nature abides in permanence as the absolute totality.

Though nature for man is *humanized*, in industry, technology, science and culture, this does not imply that nature is in general a 'social category.' Cognition of nature and its mastery is socially conditioned, and nature is a social category, changing through history, in *this* sense; but the absolute existence of nature depends on nothing and on no-one.

'If man were to transform nature entirely into an object of human, economic and productive activity, and have it cease to exist in its inviolability as nature, he would deprive himself of an essential aspect of his human life. A culture that would cull nature completely out of life would destroy itself and would become intolerable'.[27]

Man is not walled in by the subjectivity of his race, socialness, or subjective projects, in which he would merely define himself in different ways. Rather, through his being, i.e. through praxis, he has the ability to transcend his subjectivity and to get to know things as they are. The being of man reproduces not only the socio-human reality; it spiritually reproduces reality in its totality. Man exists in the totality of the world, but this totality includes man himself as well, as well as his ability to spiritually reproduce the totality of the world.

Only when man is *included* in the design of reality and when reality is grasped as the totality of nature and history will the conditions for solving the philosophical problem of man have been created. While a reality without man would be incomplete, man without the world would equally be a mere fragment. Philosophical anthropology cannot recognize the character of man for it has locked him into the subjectivity of his consciousness, race and socialness, and has radically separated him from the universe. Learning about the universe and about laws of natural events *always* also amounts to direct or indirect learning about man and his specificity.

Man's being is where the socio-human happening and the extra-human reality encounter and collide in a special way. Man is a being whose being is characterized by the practical production of the socio-human reality and by a spiritual reproduction of human and extra-human reality, of reality in general. Praxis negotiates an access to man and to comprehending man, as well as to nature and to explaining and mastering nature. The dualism of man and nature, freedom and lawfulness, anthropologism and scientism, cannot be bridged from the standpoint of consciousness or of matter, but on the basis of praxis, of praxis conceived in the above manner.

Dialectics is after the 'thing itself.' But the 'thing itself' is no ordinary thing; actually it is not a thing at all. The 'thing itself' that philosophy deals with is man and his place in the universe or, in different words: it is the totality of the world uncovered in history by man, and man existing in the totality of the world.

NOTES

[1] Descartes, *Discourse on Method*, Baltimore 1968, p. 78.

[2] In a characteristic discussion of the relation of theory and praxis, Kant demolishes the prejudices of ignoramuses who consider theory superfluous for their *fictitious* praxis. He condemns even more vehemently, however, ideas of smart-alecks that theory is good in itself but is unsuitable for praxis: 'When an ignorant individual calls theory unnecessary and dispensable in his supposed practice, this is not as unbearable as when a know-it-all admits its academic value (as a mere mental exercise, perhaps) while asserting that in practice things look altogether different'. I. Kant, *On the Old Saw: That May Be Right in Theory But It Won't Work in Practice*, Philadelphia 1974, p. 42.

[3] The connection between this new conception of reality and the genesis of modern tragedy has been pointed out by R. Grebeníčková, 'Berkovského eseje o tragédii' [Berkovski's Essays on Tragedy] in N. Berkovskij, *Eseje o tragédii* [Essays on Tragedy], Prague 1962, p. 17: 'A world in which violence reigns supreme and blood flows freely, is made of astonishingly supple material. Anything is permitted, anything can be achieved, realized, grabbed'.

[4] Husserl's distinction between theoretical and practical intentionality as well as the postulate of synthesis of universal theory and universal praxis which *changes* mankind is important in terms of the potential for further development of idealist philosophy in the 20th century.

[5] Extremely valuable historical data pertinent to this problematique have been presented in Hans Blumenberg's 'Nachahmung der Natur: Zur Vorgeschichte der Idee des schöpferischen Menschen', *Studium generale*, 1957, no. 5, pp. 266–83.

[6] Important in this context are Engels' polemical arguments: 'The subjugation of a man for menial work, in all its forms, presupposes that the subjugator has at his disposal the instruments of labor with the help of which alone he is able to employ the oppressed person and in the case of slavery, in addition, the means of subsistence which enables him to keep his slave alive'. 'Therefore, before slavery becomes possible, a certain level of production must already have been reached and a certain inequality of distribution must already have appeared'. F. Engels, *Anti-Dühring*, New York 1972, pp. 179, 178.

[7] The 'master–slave' dialectic is the basic model of praxis. Most interpreters of Hegel have missed this fundamental point.

[8] The identification of praxis in the real sense of the word with manipulation or procuring periodically leads to stressing pure theory as man's only access to the cognition of the world in its totality. Following Feuerbach, Karl Löwith also stresses that 'die alltägliche praktische Umsicht, ihr Zugriff und Angriff, versteht sich auf dieses und jenes zum Zweck der Benutzung und der Veränderung, sie erblickt aber nicht das ganze der Welt.' K. Löwith, *Gesammelte Abhandlungen*, Stuttgart 1960, p. 243. Löwith, as Feuerbach before him, runs away from the 'dirty praxis of hagglers,' which he fails to distinguish from praxis in the proper sense of the word, and embraces pure and disinterested theory.

[9] Real historical mediation, whose element is time, differs both from ideal conceptual mediation (Hegel) and from the fictitious illusory mediation of the romanticists.

[10] Materialist philosophy can therefore not accept a dualist ontology which radically distinguishes between nature as identity and history as dialectics. Such a dualist

ontology would be appropriate only if the philosophy of human reality were conceived as anthropology.

[11] This is Smith's thought quoted in context which is extremely important for comprehending the later reasoning of Kant and Hegel, far less encumbered by 'English practicism': The capitalist 'intends only his own security; and by directing . . . industry in such a manner as its produce may be of the greatest value, he intends only his own gain, and he is in this, as in many other cases, led by an invisible hand to promote an end which was no part of his intention.' A. Smith, *Wealth of Nations*, New York 1937, p. 423.

[12] Kant foreshadowed Hegel's 'cunning of reason' in 1784: 'Individuals and even whole people think little of this, that while each according to his own inclination furthers his own intention, often in opposition to others, each individual and people, as if following some guiding thread, unwittingly further the intention of Nature . . .' I. Kant, 'Idea for a Universal History.' in L. W. Beck (ed.), *Kant on History*, New York 1963, pp. 11–12 (adapted).

[13] Schelling, *Werke*, vol. 2, p. 603.

[14] The relationship of freedom and necessity is a central question for German classical philosophy. See A. F. Asmus, *Marks i burzhoaznii istorizm*, Moscow 1933, p. 68. The historical parts of this work, especially its investigation of historical and philosophical problems of Hobbes, Spinoza, Schelling and Hegel, have still not lost their scientific merit.

[15] 'What an unsophisticated perspective might accidentally consider free and thus objective is in reality predetermined and necessary, with the individual making it his *own* act. This, incidentally, is for better or worse the tool of absolute necessity, which goes for success as well'. Schelling, *Werke*, vol. 3, p. 313. The Czech Augustin Smetana ironically commented that Schelling had brought the problem to a head in his philosophy, but in solving it he lowered the flag of science and hoisted that of faith. Schelling's formulation would solve the contradiction of freedom and necessity only 'if we could delete the stamp of freedom from the concept of action, that is, if there were no contradiction in the first place.' (A. Smetana, *Sebrané spisy* [Collected Works], Prague 1960, pp. 66f.) Contemporary philosophers would agree with this position. H. Fuhrmanns, the editor of Schelling's work on freedom, characterizes the concept of freedom in Hegel's and Schelling's philosophy of history thus: 'Freedom is . . . the voluntary service to something preexisting'. Schelling, *Das Wesen der menschlichen Freiheit*, Düsseldorf 1950, p. xv. Another author has this to say of Schelling: 'Compared with the power of determinants which act subterraneously in history, the spontaneity of the individual decision does not signify much: if indeed one may ascribe any meaning at all to it'. H. Barth, *Philosophie der Erscheinung*, Basel 1959, vol. 2, pp. 269f.

[16] N. Abbagnano, *Posibilità e libertà*, Turin 1956, pp. 26f.

[17] Swiss scholars have calculated that some 3.640 million people have been killed in wars so far.

[18] Hegel criticizes the 'beautiful spirit' of the romanticists which knows that the world is dirty and does not want to soil itself by contact with it, i.e. through activity. This critique, levelled from the perspective of *historical activity*, cannot be identified with the 'critique' written by inmates of the 'human zoo' who denounce the 'beautiful spirit' only in order to cover up in 'historical' slogans their dreary private shop-keeping business where exactly nothing save the private interest of the shopkeeper is at stake.

[19] Potest igitur homo esse humanus deus atque deus humaniter, potest esse humanus angelus, humana bestia, humanus leo aut ursus, aut alius quodcumque.

[20] Cardinal Nicholas Cusanus is the author of this revolutionary anti-theological

conception: 'Non ergo activae creationis humanitatis alius extat finis quam humanitas'. See E. Cassirer, *The Individual and the Cosmos in Renaissance Philosophy*, New York 1964 p. 87.

[21] The character and size of this book do not allow us to conduct a thorough historical investigation of Marx's spiritual development. Such an investigation would, however, demonstrate that the *subject–object problem is the central point in the clash between materialist philosophy and Hegel*; we could detect and profusely illustrate how Marx dealt with this issue *both in his early stage and in the stage of 'Capital'*. Particularly enlightening as to the history of this problem is the first edition of *Das Kapital*, of 1867. Later editions left out a great part of his explicit polemics with Hegel.

[22] Three basic moments stand out in history: the dialectics of consciousness and activity; the dialectics of intentions and the results of human activity; and the dialectics of being and people's consciousness, i.e. the oscillation between what people actually are and what they consider themselves to be (and what others consider them to be), between the real and the apparent significance and character of their activity. The permeation and unity of these elements is the basis for the multidimensionality of history.

[23] M. Buber, *Das Problem des Menschen*, Heidelberg 1948, pp. 9*f.*

[24] '... at no time in his history has man been so much of a problem to himself as he is now.' M. Scheler, *Man's Place in Nature*, Boston 1961, p. 6.

[25] M. Heidegger, *Kant and the Problem of Metaphysics*, Bloomington 1962, p. 216.

[26] 'Im Grunde könnte man all dies zur Anthropologie rechnen, weil sich die drei ersten Fragen auf die letzte beziehen'. I. Kant, *Werke*, Frankfurt 1964, vol. 6, p. 448.

[27] S. L. Rubinstein, *Printsipi i put'i*, p. 205.

INDEX OF NAMES

SYNTHESE LIBRARY

Monographs on Epistemology, Logic, Methodology,
Philosophy of Science, Sociology of Science and of Knowledge, and on the
Mathematical Methods of Social and Behavioral Sciences

Managing Editor:

JAAKKO HINTIKKA (Academy of Finland and Stanford University)

Editors:

ROBERT S. COHEN (Boston University)
DONALD DAVIDSON (The Rockefeller University and Princeton University)
GABRIËL NUCHELMANS (University of Leyden)
WESLEY C. SALMON (University of Arizona)

1. J. M. BOCHENSKI, *A Precis of Mathematical Logic*. 1959, X + 100 pp.
2. P. L. GUIRAUD, *Problèmes et méthodes de la statistique linguistique*. 1960, VI + 146 pp.
3. HANS FREUDENTHAL (ed.), *The Concept and the Role of the Model in Mathematics and Natural and Social Sciences, Proceedings of a Colloquium held at Utrecht, The Netherlands, January 1960*. 1961, VI + 194 pp.
4. EVERT W. BETH, *Formal Methods. An Introduction to Symbolic Logic and the Study of effective Operations in Arithmetic and Logic*. 1962, XIV + 170 pp.
5. B. H. KAZEMIER and D. VUYSJE (eds.), *Logic and Language. Studies dedicated to Professor Rudolf Carnap on the Occasion of his Seventieth Birthday*. 1962, VI + 256 pp.
6. MARX W. WARTOFSKY (ed.), *Proceedings of the Boston Colloquium for the Philosophy of Science, 1961–1962*, Boston Studies in the Philosophy of Science (ed. by Robert S. Cohen and Marx W. Wartofsky), Volume I. 1973, VIII + 212 pp.
7. A. A. ZINOV'EV, *Philosophical Problems of Many-Valued Logic*. 1963. XIV + 155 pp.
8. GEORGES GURVITCH, *The Spectrum of Social Time*. 1964, XXVI + 152 pp.
9. PAUL LORENZEN, *Formal Logic*. 1965, VIII + 123 pp.
10. ROBERT S. COHEN and MARX W. WARTOFSKY (eds.), *In Honor of Philipp Frank*, Boston Studies in het Philosophy of Science (ed. by Robert S. Cohen and Marx W. Wartofsky), Volume II. 1965, XXXIV + 475 pp.
11. EVERT W. BETH, *Mathematical Thought. An Introduction to the Philosopy of Mathematics*. 1965, XII + 208 pp.
12. EVERT W. BETH and JEAN PIAGET, *Mathematical Epistemology and Psychology*. 1966, XII + 326 pp.
13. GUIDO KÜNG, *Ontology and the Logistic Analysis of Language. An Enquiry into the Contemporary Views on Universals*. 1967, XI + 210 pp.
14. ROBERT S. COHEN and MARX W. WARTOFSKY (eds.), *Proceedings of the Boston Colloquium for the Philosophy of Science 1964–1966, in Memory of Norwood Russell Hanson*, Boston Studies in the Philosophy of Science (ed. by Robert S. Cohen and Marx W. Wartofsky), Volume III. 1967, XLIX + 489 pp.

15. C. D. BROAD, *Induction, Probability, and Causation. Selected Papers*. 1968, XI + 296 pp.
16. GÜNTHER PATZIG, *Aristotle's Theory of the Syllogism. A logical-Philosophical Study of Book A of the Prior Analytics*. 1968, XVII + 215 pp.
17. NICHOLAS RESCHER, *Topics in Philosophical Logic*. 1968, XIV + 347 pp.
18. ROBERT S. COHEN and MARX W. WARTOFSKY (eds.), *Proceedings of the Boston Colloquium for the Philosophy of Science 1966–1968*, Boston Studies in the Philosophy of Science (ed. by Robert S. Cohen and Marx W. Wartofsky), Volume IV. 1969, VIII + 537 pp.
19. ROBERT S. COHEN and MARX W. WARTOFSKY (eds.), *Proceedings of the Boston Colloquium for the Philosophy of Science 1966–1968*, Boston Studies in the Philosophy of Science (ed. by Robert S. Cohen and Marx W. Wartofsky), Volume V. 1969, VIII + 482 pp.
20. J. W. DAVIS, D. J. HOCKNEY, and W. K. WILSON (eds.), *Philosophical Logic*. 1969, VIII + 277 pp.
21. D. DAVIDSON and J. HINTIKKA (eds.), *Words and Objections. Essays on the Work of W. V. Quine*. 1969, VIII + 366 pp.
22. PATRICK SUPPES, *Studies in the Methodology and Foundations of Science. Selected Papers from 1911 to 1969*, XII + 473 pp.
23. JAAKKO HINTIKKA, *Models for Modalities. Selected Essays*. 1969, IX + 220 pp.
24. NICHOLAS RESCHER *et al.* (eds.), *Essays in Honor of Carl G. Hempel. A Tribute on the Occasion of his Sixty-Fifth Birthday*. 1969, VII + 272 pp.
25. P.V. TAVANEC (ed.), *Problems of the Logic of Scientific Knowledge*. 1969, VII + 429 pp.
26. MARSHALL SWAIN (ed.), *Induction, Acceptance, and Rational Belief*. 1970, VII + 232 pp.
27. ROBERT S. COHEN and RAYMOND J. SEEGER (eds.), *Ernst Mach: Physicist and Philosopher*, Boston Studies in the Philosophy of Science (ed. by Robert S. Cohen and Marx W. Wartofsky), Volume VI. 1970, VIII + 295 pp.
28. JAAKKO HINTIKKA and PATRICK SUPPES, *Information and Inference*. 1970, X + 336 pp.
29. KAREL LAMBERT, *Philosophical Problems in Logic. Some Recent Developments*. 1970, VII + 176 pp.
30. ROLF A. EBERLE, *Nominalistic Systems*. 1970, IX + 217 pp.
31. PAUL WEINGARTNER and GERHARD ZECHA (eds.), *Induction, Physics, and Ethics. Proceedings and Discussions of the 1968 Salzburg Colloquium in the Philosophy of Science*. 1970, X + 382 pp.
32. EVERT W. BETH, *Aspects of Modern Logic*. 1970, XI + 176 pp.
33. RISTO HILPINEN (ed.), *Deontic Logic: Introductory and Systematic Readings*. 1971, VII + 182 pp.
34. JEAN-LOUIS KRIVINE, *Introduction to Axiomatic Set Theory*. 1971, VII + 98 pp.
35. JOSEPH D. SNEED, *The Logical Structure of Mathematical Physics*. 1971, XV + 311 pp.
36. CARL R. KORDIG, *The Justification of Scientific Change*. 1971, XIV + 119 pp.
37. MILIČ ČAPEK, *Bergson and Modern Physics*, Boston Studies in the Philosophy of Science (ed. by Robert S. Cohen and Marx W. Wartofsky), Volume VII. 1971, XV + 414 pp.
38. NORWOOD RUSSELL HANSON, *What I do Not Believe, and Other Essays* (ed. by Stephen Toulmin and Harry Woolf), 1971, XII + 390 pp.
39. ROGER C. BUCK and ROBERT S. COHEN (eds.), *PSA 1970. In Memory of Rudolf Carnap*, Boston Studies in the Philosophy of Science (ed. by Robert S. Cohen and Marx W. Wartofsky), Volume VIII. 1971, LXVI + 615 pp. Also available as paperback.
40. DONALD DAVIDSON and GILBERT HARMAN (eds.), *Semantics of Natural Language*. 1972, X + 769 pp. Also available as paperback.

41. YEHOSHUA BAR-HILLEL (ed.), *Pragmatics of Natural Languages*. 1971, VII + 231 pp.
42. SÖREN STENLUND, *Combinators, λ-Terms and Proof Theory*. 1972, 184 pp.
43. MARTIN STRAUSS, *Modern Physics and Its Philosophy. Selected Papers in the Logic, History, and Philosophy of Science*. 1972, X + 297 pp.
44. MARIO BUNGE, *Method, Model and Matter*. 1973, VII + 196 pp.
45. MARIO BUNGE, *Philosophy of Physics*. 1973, IX + 248 pp.
46. A. A. ZINOV'EV, *Foundations of the Logical Theory of Scientific Knowledge (Complex Logic)*, Boston Studies in the Philosophy of Science (ed. by Robert S. Cohen and Marx W. Wartofsky), Volume IX. Revised and enlarged English edition with an appendix, by G. A. Smirnov, E. A. Sidorenka, A. M. Fedina, and L. A. Bobrova. 1973, XXII + 301 pp. Also available as paperback.
47. LADISLAV TONDL, *Scientific Procedures*, Boston Studies in the Philosophy of Science (ed. by Robert S. Cohen and Marx W. Wartofsky), Volume X. 1973, XII + 268 pp. Also available as paperback.
48. NORWOOD RUSSELL HANSON, *Constellations and Conjectures*, (ed. by Willard C. Humphreys, Jr.), 1973, X + 282 pp.
49. K. J. J. HINTIKKA, J. M. E. MORAVCSIK, and P. SUPPES (eds.), *Approaches to Natural Language. Proceedings of the 1970 Stanford Workshop on Grammar and Semantics*. 1973, VIII + 526 pp. Also available as paperback.
50. MARIO BUNGE (ed.), *Exact Philosophy – Problems, Tools, and Goals*. 1973, X + 214 pp.
51. RADU J. BOGDAN and ILKKA NIINILUOTO (eds.), *Logic, Language, and Probability*. A selection of papers contributed to Sections IV, VI, and XI of the Fourth International Congress for Logic, Methodology, and Philosophy of Science, Bucharest, September 1971. 1973, X + 323 pp.
52. GLENN PEARCE and PATRICK MAYNARD (eds.), *Conceptual Chance*. 1973, XII + 282 pp.
53. ILKKA NIINILUOTO and RAIMO TUOMELA, *Theoretical Concepts and Hypothetico-Inductive Inference*. 1973, VII + 264 pp.
54. ROLAND FRAÏSSÉ, *Course of Mathematical Logic – Volume 1: Relation and Logical Formula*. 1973, XVI + 186 pp. Also available as paperback.
55. ADOLF GRÜNBAUM, *Philosophical Problems of Space and Time*. Second, enlarged edition, Boston Studies in the Philosophy of Science (ed. by Robert S. Cohen and Marx W. Wartofsky), Volume XII. 1973, XXIII + 884 pp. Also available as paperback.
56. PATRICK SUPPES (ed.), *Space, Time, and Geometry*. 1973, XI + 424 pp.
57. HANS KELSEN, *Essays in Legal and Moral Philosophy*, selected and introduced by Ota Weinberger. 1973, XXVIII + 300 pp.
58. R. J. SEEGER and ROBERT S. COHEN (eds.), *Philosophical Foundations of Science. Proceedings of an AAAS Program, 1969*. Boston Studies in the Philosophy of Science (ed. by Robert S. Cohen and Marx W. Wartofsky), Volume XI. 1974, X + 545 pp. Also available as paperback.
59. ROBERT S. COHEN and MARX W. WARTOFSKY (eds.), *Logical and Epistemological Studies in Contemporary Physics*, Boston Studies in the Philosophy of Science (ed. by Robert S. Cohen and Marx W. Wartofsky), Volume XIII. 1973, VIII + 462 pp. Also available as paperback.
60. ROBERT S. COHEN and MARX W. WARTOFSKY (eds.), *Methodological and Historical Essays in the Natural and Social Sciences. Proceedings of the Boston Colloquium for the Philosophy of Science, 1969–1972.* Boston Studies in the Philosophy of Science (ed. by Robert S. Cohen and Marx W. Wartofsky), Volume XIV. 1974, VIII + 405 pp. Also available as paperback.
61. ROBERT S. COHEN, J. J. STACHEL and MARX W. WARTOFSKY (eds.), *For Dirk Struik*.

Scientific, Historical and Polical Essays in Honor of Dirk J. Struik, Boston Studies in the Philosophy of Science (ed. by Robert S. Cohen and Marx W. Wartofsky), Volume XV. 1974, XXVII + 652 pp. Also available as paperback.

62. KAZIMIERZ AJDUKIEWICZ, *Pragmatic Logic*, transl. from the Polish by Olgierd Wojtasiewicz. (1974, XV + 460 pp.

63. SÖREN STENLUND (ed.), *Logical Theory and Semantic Analysis. Essays Dedicated to Stig Kanger on His Fiftieth Birthday.* 1974, V + 217 pp.

64. KENNETH F. SCHAFFNER and ROBERT S. COHEN (eds.), *Proceedings of the 1972 Biennial Meeting, Philosophy of Science Association*, Boston Studies in the Philosophy of Science (ed. by Robert S. Cohen and Marx W. Wartofsky), Volume XX. 1974, IX + 444 pp. Also available as paperback.

65. HENRY E. KYBURG, JR., *The Logical Foundations of Statistical Inference.* 1974, IX + 421 pp.

66. MARJORIE GRENE, *The Understanding of Nature: Essays in the Philosophy of Biology*, Boston Studies in the Philosophy of Science (ed. by Robert S. Cohen and Marx W. Wartofsky), Volume XXIII. 1974, XII + 360 pp. Also available as paperback.

67. JAN M. BROEKMAN, *Stucturalism: Moscow, Prague, Paris.* 1974, IX + 117 pp.

68. NORMAN GESCHWIND, *Selected Papers on Language and the Brain*, Boston Studies in the Philosophy of Science (ed. by Robert S. Cohen and Marx W. Wartofsky), Volume XVI. 1974, XII + 549 pp. Also available as paperback.

69. ROLAND FRAÏSSÉ, *Course of Mathematical Logic* – Volume II: *Model Theory.* 1974, XIX + 192 pp.

70. ANDRZEJ GRZEGORCZYK, *An Outline of Methematical Logic. Fundamental Results and Notions Explained with All Details.* 1974, X + 596 pp.

71. FRANZ VON KUTSCHERA, *Philosophy of Language.* 1975, VII + 305 pp.

72. JUHA MANNINEN and RAIMO TUOMELA (eds.), *Essays on Explanation and Understanding. Studies in the Foundations of Humanities and Social Sciences.* 1976, VII + 440 pp.

73. JAAKKO HINTIKKA (ed.), *Rudolf Carnap, Logical Empiricist. Materials and Perspectives.* 1975, LXVIII + 400 pp.

74. MILIČ ČAPEK (ed.), *The Concepts of Space and Time. Their Structure and Their Development.* Boston Studies in the Philosophy of Science (ed. by Robert S. Cohen and Marx W. Wartofsky), Volume XXII. 1976, LVI + 570 pp. Also available as paperback.

75. JAAKKO HINTIKKA and UNTO REMES, *The Method of Analysis. Its Geometrical Origin and Its General Significance.* Boston Studies in the Philosophy of Science (ed. by Robert S. Cohen and Marx W. Wartofsky), Volume XXV. 1974, XVIII + 144 pp. Also available as paperback.

76. JOHN EMERY MURDOCH and EDITH DUDLEY SYLLA, *The Cultural Context of Medieval Learning. Proceedings of the First International Colloquium on Philosophy, Science, and Theology in the Middle Ages – September 1973.* Boston Studies in the Philosophy of Science (ed. by Robert S. Cohen and Marx W. Wartofsky), Volume XXVI. 1975, X + 566 pp. Also available as paperback.

77. STEFAN AMSTERDAMSKI, *Between Experience and Metaphysics. Philosophical Problems of the Evolution of Science.* Boston Studies in the Philosophy of Science (ed. by Robert S. Cohen and Marx W. Wartofsky), Volume XXXV. 1975, XVIII + 193 pp. Also available as paperback.

78. PATRICK SUPPES (ed.), *Logic and Probability in Quantum Mechanics.* 1976, XV + 541 pp.

80. JOSEPH AGASSI, *Science in Flux.* Boston Studies in the Philosophy of Science (ed. by Robert S. Cohen and Marx W. Wartofsky), Volume XXVIII. 1975, XXVI + 553 pp. Also available as paperback.

81. SANDRA G. HARDING (ed.), *Can Theories Be Refuted? Essays on the Duhem-Quine Thesis*. 1976, XXI + 318 pp. Also available as paperback.

84. MARJORIE GRENE and EVERETT MENDELSOHN (eds.), *Topics in the Philosophy of Biology*. Boston Studies in the Philosophy of Science (ed. by Robert S. Cohen and Marx W. Wartofsky), Volume XXVII. 1976, XIII + 454 pp. Also available as paperback.

85. E. FISCHBEIN, *The Intuitive Sources of Probabilistic Thinking in Children*. 1975, XIII + 204 pp.

86. ERNEST W. ADAMS, *The Logic of Conditionals. An Application of Probability to Deductive Logic*. 1975, XIII + 156 pp.

89. A. KASHER (ed.), *Language in Focus: Foundations, Methods and Systems. Essays Dedicated to Yehoshua Bar-Hillel*. Boston Studies in the Philosophy of Science (ed. by Robert S. Cohen and Marx W. Wartofsky), Volume XLIII. 1976, XXVIII + 679 pp. Also available as paperback.

90. JAAKKO HINTIKKA, *The Intentions of Intentionality and Other New Models for Modalities*. 1975, XVIII + 262 pp. Also available as paperback.

93. RADU J. BOGDAN, *Local Induction*. 1976, XIV + 340 pp.

95. PETER MITTELSTAEDT, *Philosophical Problems of Modern Physics*. Boston Studies in the Philosophy of Science (ed. by Robert S. Cohen and Marx W. Wartofsky), Volume XVIII. 1976, X + 211 pp. Also available as paperback.

96. GERALD HOLTON and WILLIAM BLANPIED (eds.), *Science and Its Public: The Changing Relationship*. Boston Studies in the Philosophy of Science (ed. by Robert S. Cohen and Marx W. Wartofsky), Volume XXXIII. 1976, XXV + 289 pp. Also available as paperback.

97. MYLES BRAND and DOUGLAS WALTON (eds.), *Action Theory. Proceedings of the Winnipeg Conference on Human Action, Held at Winnipeg, Manitoba, Canada, 9-11 May 1975*. 1976, VI + 345 pp.

SYNTHESE HISTORICAL LIBRARY

Texts and Studies
in the History of Logic and Philosophy

Editors:

N. KRETZMANN (Cornell University)
G. NUCHELMANS (University of Leyden)
L. M. DE RIJK (University of Leyden)

1. M. T. BEONIO-BROCCHIERI FUMAGALLI, *The Logic of Abelard.* Translated from the Italian. 1969, IX + 101 pp.
2. GOTTFRIED WILHELM LEIBNITZ, *Philosophical Papers and Letters.* A selection translated and edited, with an introduction, by Leroy E. Loemker. 1969, XII + 736 pp.
3. ERNST MALLY, *Logische Schriften,* ed. by Karl Wolf and Paul Weingartner. 1971, X + 340 pp.
4. LEWIS WHITE BECK (ed.), *Proceedings of the Third International Kant Congress.* 1972, XI + 718 pp.
5. BERNARD BOLZANO, *Theory of Science,* ed. by Jan Berg. 1973, XV + 398 pp.
6. J. M. E. MORAVCSIK (ed.), *Patterns in Plato's Thought. Papers arising out of the 1971 West Coast Greek Philosophy Conference.* 1973, VIII + 212 pp.
7. NABIL SHEHABY, *The Propositional Logic of Avicenna: A Translation from al-Shifā: al-Qiyās,* with Introduction, Commentary and Glossary. 1973, XIII + 296 pp.
8. DESMOND PAUL HENRY, *Commentary on De Grammatico: The Historical-Logical Dimensions of a Dialogue of St. Anselm's.* 1974, IX + 345 pp.
9. JOHN CORCORAN, *Ancient Logic and Its Modern Interpretations.* 1974, X + 208 pp.
10. E. M. BARTH, *The Logic of the Articles in Traditional Philosophy.* 1974, XXVII + 533 pp.
11. JAAKKO HINTIKKA, *Knowledge and the Known. Historical Perspectives in Epistemology.* 1974, XII + 243 pp.
12. E. J. ASHWORTH, *Language and Logic in the Post-Medieval Period.* 1974, XIII + 304 pp.
13. ARISTOTLE, *The Nicomachean Ethics.* Translated with Commentaries and Glossary by Hypocrates G. Apostle. 1975, XXI + 372 pp.
14. R. M. DANCY, *Sense and Contradiction: A Study in Aristotle.* 1975, XII + 184 pp.
15. WILBUR RICHARD KNORR, *The Evolution of the Euclidean Elements. A Study of the Theory of Incommensurable Magnitudes and Its Significance for Early Greek Geometry.* 1975, IX + 374 pp.
16. AUGUSTINE, *De Dialectica.* Translated with the Introduction and Notes by B. Darrell Jackson. 1975, XI + 151 pp.